Applied Probability and Statistics (Continued)

BROWNLEE · Statistical Theory and Me Engineering, *Second Edition*

CHAKRAVARTI, LAHA and ROY · Handbook of Methods of Applied Statistics, Vol. I

CHAKRAVARTI, LAHA and ROY · Handbook of Methods of Applied Statistics, Vol. II

CHERNOFF and MOSES · Elementary Decision Theory

CHIANG · Introduction to Stochastic Processes in Biostatistics

CLELLAND, deCANI, BROWN, BURSK, and MURRAY · Basic Statistics with Business Applications

COCHRAN · Sampling Techniques, *Second Edition*

COCHRAN and COX · Experimental Designs, *Second Edition*

COX · Planning of Experiments

COX and MILLER · The Theory of Stochastic Processes

DANIEL and WOOD · Fitting Equations to Data

DAVID · Order Statistics

DEMING · Sample Design in Business Research

DODGE and ROMIG · Sampling Inspection Tables, *Second Edition*

DRAPER and SMITH · Applied Regression Analysis

ELANDT-JOHNSON · Probability Models and Statistical Methods in Genetics

GOLDBERGER · Econometric Theory

GUTTMAN, WILKS and HUNTER · Introductory Engineering Statistics, *Second Edition*

HAHN and SHAPIRO · Statistical Models in Engineering

HALD · Statistical Tables and Formulas

HALD · Statistical Theory with Engineering Applications

HOEL · Elementary Statistics, *Third Edition*

HUANG · Regression and Econometric Methods

JOHNSON and KOTZ · Distributions in Statistics
- Discrete Distributions
- Continuous Univariate Distributions-1
- Continuous Univariate Distributions-2
- Continuous Multivariate Distributions

JOHNSON and LEONE · Statistics and Experimental Design: In Engineering and the Physical Sciences, Volumes I and II

LANCASTER · The Chi Squared Distribution

LEWIS · Stochastic Point Processes

MILTON · Rank Order Probabilities: Two-Sample Normal Shift Alternatives

RAO and MITRA · Generalized Inverse of Matrices and Its Applications

SARD and WEINTRAUB · A Book of Splines

SARHAN and GREENBERG · Contributions to Order Statistics

SEAL · Stochastic Theory of a Risk Business

SEARLE · Linear Models

THOMAS · An Introduction to Applied Probability and Random Processes

WHITTLE · Optimization under Constraints

WILLIAMS · Regression Analysis

continued on back

Distributions

in Statistics

A WILEY PUBLICATION IN APPLIED STATISTICS

Distributions in Statistics: Continuous Multivariate Distributions

NORMAN L. JOHNSON
University of North Carolina, Chapel Hill

SAMUEL KOTZ
Temple University, Philadelphia

John Wiley & Sons, Inc.
New York . London . Sydney . Toronto

Library of Congress Cataloging in Publication Data

Johnson, Norman Lloyd.
Continuous multivariate distributions.

(Their Distributions in statistics) (A Wiley publication in applied statistics) (Wiley series in probability and mathematical statistics)
Includes bibliographies.
1. Distribution (Probability theory) 2. Multivariate analysis. I. Kotz, Samuel, joint author. II. Title.

QA273.6.J58 519.5'3 72-1342
ISBN 0-471-44370-0

Printed in the United States of America

10 9 8 7 6 5 4 3 2 1

To the memory of
Dr. Pauline Kotz-Zelikovsky

Preface

When working with multivariate distributions, it is almost essential to use matrices to present formulas in a compact way. In this book matrices (including vectors) are denoted by boldface type. Random variables are usually (but not always) given in capital letters.

The distinction between modeling and sampling distributions, referred to in the Preface to the volumes on univariate continuous distributions, is even more marked with multivariate distributions. Sampling distributions (mostly of statistics obtained from random samples from multinormal populations) are concentrated in Chapters 38 and 39. They are generally much more complicated in mathematical expression than are the modeling distributions. In Chapter 38 we introduce a number of mathematical functions that greatly simplify the formal representation of many sampling distributions, though this hides quite formidable difficulties in numerical computation.

It will be noted that in the later chapters there is an absence of historical remarks. The main reason is that the topics discussed have relatively little history. Much multivariate theory has become practically useful only with the advent of the electronic computer. This stimulus has resulted in a rapid development of multivariate theory over the past two decades. Much of this work is still not complete (adequate tables of multivariate t, for example, are only now becoming available), and we think we are too close to it to attempt a summary historical assessment.

It is noteworthy that, among nonnormal distributions, methods of estimation appear to have been developed less fully than one might expect. We have, therefore, in Chapter 42, devoted a rather large amount of space to estimation of a particular parameter in the beta-Stacy distribution, as an indication of the kinds of problems that may arise in other settings.

We should like to express our thanks to Mr. P. Chen, Mr. S. Kulpinsky, and Mrs. J. Oler for help in the initial work on Chapters 35 and 36 and to Dr. P. Kanofsky for similar help on Chapter 37. We are grateful as well to Mrs. C. Tuker and Mrs. M. Hopkins, librarians at Temple University and the University of North Carolina at Chapel Hill, and to our patient and hardworking typists, Mrs. G. Ballard and Mrs. Susan Drum.

Dr. G. S. Watson and Dr. J. W. Pratt made many useful comments on an early version of the book which resulted in substantial improvements.

We should also like to acknowledge the hospitality of the University of New South Wales, Australia, Pennsylvania State University, and Bar-Ilan University, Israel, at which parts of this work were completed. We further express our appreciation of the help and encouragement of our wives, Regina and Rosalie.

We thank Beatrice Shube for her skillful handling far beyond the call of duty of editorial work connected with this volume and Marcia Heim for her patient and efficient management of production matters.

As in the earlier volumes in this series, we acknowledge our indebtedness to correspondents in all parts of the globe who have been ready to help us when requested. We should particularly like to mention Dr. P. R. Krishnaiah, whole help we solicited on several occasions and who always gave a generous response.

NORMAN L. JOHNSON
SAMUEL KOTZ

Chapel Hill, North Carolina
Philadelphia, Pennsylvania

April 1972

A Note to the Reader

This is the fourth, and final, volume of a series on Distributions in Statistics. The first three volumes are:

Volume 1: *Discrete Distributions*
Volume 2: *Continuous Univariate Distributions 1*
Volume 3: *Continuous Univariate Distributions 2*

(Multivariate discrete distributions are discussed in Chapter 11 of Volume 1.) The present volume is intended to cover material published up to the end of 1971.

The Augmented Cumulative Index at the end of this volume covers all four volumes. For each of the first three volumes it includes a considerable additional number of entries, beyond those given in the separate indexes for each volume. In particular, we have included entries relating to "associations" between different distributions (denoted by the symbol $\leftrightarrow$).

There is also an Index of References by Topics classified according to the subject. This, again, covers all four volumes. The entries are reference numbers, preceded by chapter number.

Chapters are numbered sequentially through the four volumes:

Volume 1: Chapters 1 to 11
Volume 2: Chapters 12 to 21
Volume 3: Chapters 22 to 33
Volume 4: Chapters 34 to 42

(A list of correction to the first 3 volumes is included on pages 328–330.)

At the conclusion of this work, we would like to record our growing belief in the continued relevance (especially in multivariate analysis) of the statement made in 1951 by M. G. Kendall: "Much of present-day statistics is essentially a study of distribution functions."

N. L. J.
S. K.

Contents

Distributions
in Statistics

34

Systems of Multivariate Continuous Distributions

1. Introduction

The multinormal distribution (which will be the subject of Chapter 35) has been studied far more extensively than any other multivariate distribution. Indeed, its position of preeminence among multivariate continuous distributions is more marked than that of the normal among univariate continuous distributions. However, in recent years there have been signs that the need for usable alternatives to the multinormal distribution is becoming recognized. In the present chapter we describe some systems of multivariate distributions that may provide acceptable models for practical use.

There is some parallelism between this chapter and Chapter 12. In particular, we shall describe systems of distributions based on

(a) generalizations of Pearson's differential equation (Section 4.1, Chapter 12)

(b) series expansions—especially Gram-Charlier and Edgeworth expansions (Section 4.2, Chapter 12) and

(c) transformation to multinormal joint distributions (Section 4.3, Chapter 12).

Basic concepts relevant to multivariate distributions have been introduced in Chapters 1 and 12. Some essential features distinguishing multivariate from univariate studies have also been mentioned in Chapter 11. Correlation and regression, which are among these features, appear throughout this volume with some regularity. Here we introduce three new functions—the *scedastic*, *clisy* (or *clitic*), and *kurtic* functions of a random variable Y given the values $x_1,x_2,\ldots,x_s$ of random variables $X_1,X_2,\ldots,X_s$, which are

$$\text{var}(Y \mid x_1,x_2,\ldots,x_s),$$

$$\alpha_3(Y \mid x_1,x_2,\ldots,x_s),$$

and

$$\alpha_4(Y \mid x_1,x_2,\ldots,x_s).$$

respectively. We recall [(117) of Chapter 1] the definition of central product moments. The $m \times m$ matrix with (i,j)th element equal to the covariance of X_i and X_j [and (i,i)th element equal to the variance of X_i] is called the *variance-covariance* matrix of $\mathbf{X}' = (X_1,X_2,\ldots,X_m)$—sometimes written $\mathbf{Var}(\mathbf{X})$.

Mixtures of multivariate distributions are formed as for univariate distributions (see Chapter 1, Section 7.3). If $X_1,\ldots,X_m$ have a joint distribution which is a mixture of k distributions with cumulative distribution functions $\{F_j(x_1,\ldots,x_m)\}$, with weights $\{a_j\}$

$$\left(j = 1,\ldots,k; a_j > 0; \sum_{j=1}^{k} a_j = 1\right),$$

then

$$F_{X_1\cdots X_m}(x_1,\ldots,x_m) = \sum_{j=1}^{k} a_j F_j(x_1,\ldots,x_m). \tag{1}$$

The joint distribution of any subset of the X's is also a mixture with k components, and the same weights $\{a_j\}$. In particular

$$F_{X_1\cdots X_s}(x_1,\ldots,x_s) = \sum_{j=1}^{k} a_j F_j(x_1,\ldots,x_s) \tag{2.1}$$

and

$$F_{X_1}(x_1) = \sum_{j=1}^{k} a_j F_j(x_1), \tag{2.2}$$

where

$$F_j(x_1,\ldots,x_s) = \lim_{\substack{x_i \to \infty \\ i > s}} F_j(x_1,\ldots,x_m).$$

Most of the multivariate distributions encountered in this volume are purely continuous. There is a notable exception in Chapter 41, Section 3.4, and it should be noted that it is possible for multivariate distributions to be of mixed type, even when each marginal distribution is continuous. Some interesting examples are given by Koopmans [34].

The concept of exchangeability is of some importance. Variables $X_1, X_2, \ldots, X_m$ are said to be *exchangeable* if

$$\Pr\left[\bigcap_{j=1}^{m} (X_j \le x_j)\right] = \Pr\left[\bigcap_{j=1}^{m} (X_j \le x_{a_j})\right], \tag{3}$$

where $(a_1, \ldots, a_m)$ is any permutation of the integers $(1, \ldots, m)$.

The joint distribution is also, rather inappropriately, sometimes called "exchangeable" in these circumstances. A better term, used by Lancaster [39], is "symmetrical." Necessary and sufficient conditions are given for a distribution to be symmetrical in the first part of Section 4 below.

The "characteristic coefficients" of distributions constructed by Sarmonov (see Chapter 1, Section 5) (see also [69]) have been extended to bivariate distributions by Abazaliev [1]. The characteristic coefficients of the joint distribution of X_1 and X_2 are

$$\lambda_{g_1,g_2}(r_1,r_2;X_1,X_2) = E\left[\exp\left\{2\pi i \sum_{j=1}^{2} r_j[g_j(X_j) - \tfrac{1}{2}]\right\}\right], \tag{4}$$

where $g_j(y)$ is an increasing function of y with $\lim_{y\to-\infty} g_j(y) = 0$, $\lim_{y\to\infty} g_j(y) = 1$. If X_1 and X_2 are mutually independent,

$$\lambda_{g_1,g_2}(r_1,r_2;X_1,X_2) = \prod_{j=1}^{2} \lambda_{g_j}(r_j;X_j),$$

where $\lambda_{g_j}(r;X_j)$ is a characteristic coefficient for the distribution of X_j, as defined by Sarmanov [69].

In particular, $g_j(x_j)$ may be taken as the cumulative distribution function of $X_j (j = 1,2)$. Then

$$F_{X_1,X_2}(x_1,x_2) = F_{X_1}(x_1)F_{X_2}(x_2) + \pi^{-1} \sum_{r_1=1}^{\infty} \sum_{r_2=1}^{\infty} (r_1 r_2)^{-1}\lambda(r_1,r_2) \sin r_1 z_1 \sin r_2 z_2,$$

where $\lambda(r_1,r_2)$ is an abbreviation for (4), and $z_i = 2(F_{X_j}(x_j) - \frac{1}{2})$.

Abazaliev [1] shows that any bivariate distribution is determined by the coefficients (4), if the functions $g_1(\cdot)$ and $g_2(\cdot)$ are known. He further shows that if the series

$$\sum_{r_1=1}^{\infty} \sum_{r_2=1}^{\infty} (r_1 r_2)^{-1}\lambda(r_1,r_2)$$

converges, then the joint distribution is continuous.

2. Historical Remarks

Although the bivariate normal distribution (see Chapters 35 and 36) had been studied at the beginning of the nineteenth century, interest in multivariate distributions remained at a low level until it was stimulated by the work of Galton [22] in the last quarter of the century. He did not, himself, introduce new forms of joint distribution, but he developed the idea of correlation and regression and focused attention on the need for greater knowledge of possible forms of multivariate distribution.

Investigation of nonnormal joint distributions, or "skew frequency surfaces" (as nonsymmetrical forms have been termed) has generally followed lines suggested by previous work on univariate distributions. Early work in this field followed rather different lines, but was not very successful. Karl Pearson [56], whose first investigations appear to have been prompted by noting distinctly nonnormal properties of some observed joint distributions, initially tried to proceed by an analogy with the bivariate normal surface. For this distribution (see Chapter 36, Section 1) it is possible to replace a pair of correlated variables by a pair of independent ones, using a transformation corresponding to a rotation of axes. Pearson attempted to construct general systems for which this property holds. He found, however, that this method was unpromising. In fact, in general, the property cannot hold since, for independence, the rotation must produce uncorrelated variables, but this is not sufficient to ensure independence.

Pearson [57, 58] and, later, Neyman [53] also considered methods of construction of joint distributions, starting from certain requirements on the regression and scedastic functions. This was an extension of work initiated by Yule [83], who showed that assuming multiple linear regression (i.e., $E[Y \mid x_1,x_2,\ldots,x_s]$ to be a linear function of $x_1,x_2,\ldots,x_s$), the multiple regression function obtained by the method of least squares is identical with that of a multinormal distribution. Although some useful results, not requiring detailed knowledge of the actual form of distribution, were obtained by this method, calculation of derived probabilities was not usually sufficiently precise for practical purposes. Narumi [51] used the stronger requirement that the shape of each conditional ("array") distribution of one variable given the others, should be the same for all values of the conditioning variables. He also placed requirements on the "median regression" function μ (the median of Y, given $X_1 = x_1,\ldots,X_s = x_s$). In this way he did construct some definite distributions. However, the requirement of unchanging shape for the conditional distribution of Y given $x_1,\ldots,x_s$ was clearly not in accord with features of many observed distributions. One distribution that can be constructed in this way (although this was not the way in which the author, in fact, approached it) is the *Rhodes' distribution* (Rhodes [61]). Let X_1,X_2 be independent gamma variables with

$$p_{X_j}(x_j) = \frac{1}{\delta_j^{\alpha_j}\Gamma(\alpha_j)}\, x_j^{\alpha_j-1}e^{-x_j/\delta_j} \qquad (0 < x_j;\quad j = 1,2).$$

Then the joint density function of Y_1 and Y_2, where

$$(5.1) \qquad X_1 = 1 - a_1^{-1}Y_1 + a_2^{-1}Y_2; \qquad X_2 = 1 - a_2'^{-1}Y_2 + a_1'^{-1}Y_1$$

is

$$(5.2) \quad p_{Y_1,Y_2}(y_1,y_2) = \frac{e^{-(\delta_1^{-1}+\delta_2^{-1})}}{\delta_1^{\alpha_1}\delta_2^{\alpha_2}\Gamma(\alpha_1)\Gamma(\alpha_2)}\left|\frac{a_1'a_2 - a_1a_2'}{a_1a_2a_1'a_2'}\right|$$

$$\times (1 - a_1^{-1}y_1 + a_2^{-1}y_2)^{\alpha_1-1} (1 - a_2'^{-1}y_2 + a_1'^{-1}y_1)^{\alpha_2-1}e^{-\lambda_1y_1-\lambda_2y_2}$$

with

$$1 - a_1^{-1}y_1 + a_2^{-1}y_2 > 0; \quad 1 - a_2'^{-1}y_2 + a_1'^{-1}y_1 > 0.$$

Here $\lambda_1 = (a_1\delta_1)^{-1} - (a_1'\delta_2)^{-1}$; $\lambda_2 = (a_2'\delta_2)^{-1} - (a_2\delta_1)^{-1}$. The special case when $\delta_j = (\alpha_j - 1)^{-1}$ is of particular interest being the form originally proposed by Rhodes (see also Mardia [45]).

Inverting the (linear) transformation (5.1), we obtain

$$(6.1) \qquad Y_1 = \{a_2(X_1 - 1) + a_2'(X_2 - 1)\}/(a_1'^{-1}a_2'^{-1} - a_1^{-1}a_2^{-1});$$

$$(6.2) \qquad Y_2 = \{a_1(X_1 - 1) + a_1'(X_2 - 1)\}/(a_1^{-1}a_2^{-1} - a_1'^{-1}a_2'^{-1}).$$

Hence

$$\operatorname{var}(Y_1) = \frac{1}{K^2}[a_2^2\delta_1^2\alpha_1 + a_2'^2\delta_2^2\alpha_2],$$

$$\operatorname{var}(Y_2) = \frac{1}{L^2}[a_1^2\delta_1^2\alpha_1 + a_1'\delta_2^2\alpha_2],$$

$$\operatorname{cov}(Y_1,Y_2) = \frac{1}{KL}[a_1a_2\delta_1^2\alpha_1 + a_1'a_2'\delta_2^2\alpha_2],$$

and

$$\operatorname{corr}(Y_1,Y_2) = \frac{a_1a_2\delta_1^2\alpha_1 + a_1'a_2'\delta_2^2\alpha_2}{\sqrt{(a_2^2\delta_1^2\alpha_1 + a_2'^2\delta_2^2\alpha_2)(a_1^2\delta_1^2\alpha_1 + a_1'^2\delta_2^2\alpha_2)}},$$

where K and L are the denominators in (6.1) and (6.2) respectively.

Multivariate extension of Gram-Charlier and Edgeworth series expansions is the subject of Section 4 of this chapter. Work on these forms of distribution appears to have commenced rather suddenly about 1910 and continued, with slowly decreasing intensity after 1920, till Pretorius [60] gave a comprehensive survey of results available in 1930. Since that time, interest has continued at a steady, but rather low, level, with a recent increase exemplified by a paper of some generality by Chambers [10].

In Chapter 12 (Section 4.3) we have already discussed systems of distributions constructed by supposing certain (fairly simple) functions of variables to be normally distributed. It is natural to consider what forms of joint distribution one can construct by supposing certain functions of the original variables to have a joint multinormal distribution. Although, in the general case, we should consider situations where $Z_i = g_i(X_1,X_2,\ldots,X_s)$ $(i = 1,\ldots,s)$ have a joint multinormal distribution, we shall, in Section 5, consider only those cases in which $Z_i = g_i(X_i)$, i.e., when each of the original variables $(X_1,\ldots,X_s)$ is transformed separately to a normal variable.

Edgeworth [17, 18] used cubic polynomial transformations for each of two variables separately. He also considered composite polynomial transformations. Wicksell [81, 82] supposed $\log X_1$ and $\log X_2$ to have a bivariate normal distribution (the *logarithmic surface*). This distribution is discussed in Section 5 as is the semilogarithmic surface in which X_1 and $\log X_2$ have a bivariate normal distribution (Jørgensen [31]). Recent work has tended to aim at building up multivariate distributions having specified structures. These are discussed in Section 6.

3. Multivariate Generalization of Pearson System

The univariate Pearson system of distributions has been discussed in Chapter 12 (Section 4.1). Successes achieved, using these distributions, led to attempts to extend them to multivariate (in particular, bivariate) distributions. Clearly, there can be considerable variety in possible bivariate distributions with each of the marginal distributions being one or the other of the Pearson Types. However, it is reasonable to restrict ourselves to consideration of systems derived from differential equations which are natural generalizations of (23) in Chapter 12.

We describe here some investigations reported by van Uven [75–77]. These are not the only studies of this kind (see, e.g., Risser [62–64] and Risser and Traynard [65]), but they are the most exhaustive and systematic known to us. We start from the pair of differential equations

$$\frac{\partial \log p}{\partial x_j} = \frac{L_j(x_1,x_2)}{Q_j(x_1,x_2)} \qquad (j = 1,2), \tag{7}$$

where $p \equiv p_{X_1,X_2}(x_1,x_2)$ is the joint probability density function of X_1 and X_2, and L_j,Q_j are linear and quadratic functions, respectively, of their arguments. On fixing either x_1 or x_2, it is clear that the conditional ("array") distributions of either variable, given the other, satisfy differential equations of the form (23) of Chapter 12, hence belong to the Pearson system. However, since the values of the constants depend on the value of the conditioning variable, the array distributions do not, in general, all have the same shape.

From (7), we see that

$$\frac{\partial^2 \log p}{\partial x_1\, \partial x_2} = \frac{\partial}{\partial x_1}\left(\frac{L_2(x_1,x_2)}{Q_2(x_1,x_2)}\right) = Q_2^{-2}\left(Q_2\frac{\partial L_2}{\partial x_1} - L_2\frac{\partial Q_2}{\partial x_1}\right).$$

(Arguments (x_1,x_2) have been omitted for convenience.) But also

$$\frac{\partial^2 \log p}{\partial x_1\, \partial x_2} = \frac{\partial}{\partial x_2}\left(\frac{L_1}{Q_1}\right) = Q_1^{-1}\left(Q_1\frac{\partial L_1}{\partial x_2} - L_1\frac{\partial Q_1}{\partial x_2}\right).$$

Hence

$$Q_1^{-2}\left(Q_1\frac{\partial L_1}{\partial x_2} - L_1\frac{\partial Q_1}{\partial x_2}\right) = Q_2^{-2}\left(Q_2\frac{\partial L_2}{\partial x_1} - L_2\frac{\partial Q_2}{\partial x_1}\right),$$

showing that the L_j's and Q_j's cannot be chosen in a completely arbitrary manner.

From (7), with $j = 1$, integrating over the range of variation of x_1, we find

$$\int_{-\infty}^{\infty} L_1 p\, dx_1 = \int_{-\infty}^{\infty} Q_1 \frac{\partial p}{\partial x_1}\, dx_1 = [Q_1 p]_{-\infty}^{\infty} - \int_{-\infty}^{\infty} \frac{\partial Q_1}{\partial x_1}\, p\, dx_1,$$

so that

$$\int_{-\infty}^{\infty}\left(L_1 + \frac{\partial Q_1}{\partial x_1}\right) p\, dx_1 = [Q_1 p]_{-\infty}^{\infty}. \tag{8}$$

The quantity on the right-hand side of (8) is to be calculated as

$$\lim_{x_1\to\infty} Q_1 p - \lim_{x_1\to-\infty} Q_1 p.$$

Very often each of these limits is zero (whenever $\lim_{x_1\to\infty} x_1^2 p = 0 = \lim_{x_1\to-\infty} x_1^2 p$, in fact). Since L_1 and $\partial Q_1/\partial x_1$ are each linear functions of x_1 and x_2, it follows that (8) can be written in the form

$$E[X_1 \mid x_2] = \alpha_1 + \beta_1 x_2$$

if $[Q_1 p]_{-\infty}^{\infty} = 0$. This means that the regression of X_1 on X_2 is linear. Similarly, the regression of X_2 on X_1 is linear if $[Q_2 p]_{-\infty}^{\infty} = 0$.

Note that if, for given $X_2 = x_2$, the range of X_1 is finite, $a_2(x_2) < X_1 < a_1(x_2)$, then

$$[Q_1 p]_{-\infty}^{\infty} = \lim_{x_1\to a_1(x_2)} Q_1 p - \lim_{x_1\to a_2(x_2)} Q_1 p.$$

The condition $[Q_1 p]_{-\infty}^{\infty} = 0$ is satisfied if $\lim x^2 p$ is zero at each end of the range of variation.

Table 1 (reproduced from Elderton and Johnson [19]) gives the more important members of this system of bivariate Pearson distributions. They all have linear regression of either variable on the other. Most of these

TABLE 1

Bivariate Pearson Surfaces

Type	Equation $y =$	Conditions	Marginal types x_1	x_2	Chapter number
I	$f(x_1)f(x_2)$	(Independent variables with frequencies $f(x_1)$, $f(x_2)$)			
II$a\alpha$	$\frac{\Gamma(m_1 + m_2 + m_3)}{\Gamma(m_1)\Gamma(m_2)\Gamma(m_3)} x_1^{m_1-1}x_2^{m_2-1}(1 - x_1 - x_2)^{m_3-1}$	$m_1,m_2,m_3 > 0$ $x_1,x_2 > 0;\ x_1 + x_2 \leq 1$	I or II	I or II	40
II$a\beta$	$\frac{\Gamma(-m_3 + 1)x_1^{m_1-1}x_2^{m_2-1}(1 + x_1 + x_2)^{m_3-1}}{\Gamma(m_1)\Gamma(m_2)\Gamma(-m_1 - m_2 - m_3 + 1)}$	$m_1,m_2 > 0;\ m_1 + m_2 + m_3 < 0$ $x_1,x_2 > 0$	VI	VI	40
II$a\gamma$	$\frac{\Gamma(-m_2 + 1)x_1^{m_1-1}x_2^{m_2-1}(-1 - x_1 + x_2)^{m_3-1}}{\Gamma(m_1)\Gamma m_3)\Gamma(-m_1 - m_2 - m_3 + 1)}$	$m_1,m_3 > 0;\ m_1 + m_2 + m_3 < 0$ $x_2 - 1 > x_1 > 0$	VI	VI	40
II$a\delta$	$\frac{\Gamma(-m_1 + 1)x_1^{m_1-1}x_2^{m_2-1}(-1 + x_1 - x_2)^{m_3-1}}{\Gamma(m_2)\Gamma(m_3)\Gamma(-m_1 - m_2 - m_3 + 1)}$	$m_2,m_3 > 0;\ m_1 + m_2 + m_3 < 0$ $x_1 - 1 > x_2 > 0$	VI	VI	40
IIb	$\frac{x_1^{m_1-1}x_2^{m_2-1}\exp[-(x_1 + 1)/x_2]}{\Gamma(m_1)\Gamma(-m_1 - m_2)}$	$m_1 > 0;\ m_1 + m_2 < 0$ $x_1,x_2 > 0$	VI	V	41
III$a\alpha$	$\frac{-m\sqrt{(1 - \rho^2)}}{\pi k^m}(k + x_1^2 + 2\rho x_1x_2 + x_2^2)^{m-1}$	$m < 0;\ \lvert\rho\rvert < 1;\ k > 0$	VII	VII	37
III$a\beta$	$\frac{m\sqrt{(1 - \rho^2)}}{\pi k^m}(k - x_1^2 + 2\rho x_1x_2 - x_2^2)^{m-1}$	$m > 0;\ \lvert\rho\rvert < 1;\ k > 0$ $x_1^2 - 2\rho x_1x_2 + x_2^2 < k$	II	II	40
IVa	$\frac{x_1^{m_1-1}(x_2 - x_1)^{m_2-1}e^{-x_2}}{\Gamma(m_1)\Gamma(m_2)}$	$m_1,m_2 > 0$ $0 < x_1 < x_2$	III	III	41
VI	$\frac{1}{2\pi\sqrt{(1 - \rho^2)}}\exp\left[-\frac{1}{2(1 - \rho^2)}(x_1^2 - 2\rho x_1x_2 + x_2^2)\right]$	$\lvert\rho\rvert < 1$	Normal	Normal	36

distributions are treated in detail in later chapters. Where this is so, the appropriate chapter number is given in the last column of the table. (Two-dimensional extensions of the Pearson system have also been discussed by Sagrista [67].)

Steyn [74] has extended this kind of analysis to joint distributions of more than two variables. Starting with a set of m equations obtained by letting j, in (7), run from 1 to m, he showed that if p vanishes at the extremes of the range of variation of x_i, the regression of X_i on the other $(m - 1)$ variables is linear.

4. Series Expansions and Multivariate Central Limit Theorems

Some interesting general results on expansions of multivariate density functions have been obtained by Lancaster [38] (see also [40]). For continuous distributions some of these results can be expressed in relatively simple form. We first need to introduce the concept of an *orthonormal set of functions* on the distribution of a random variable X. These are simply a infinite sequence of functions $\{X_{(j)}\}$ of X such that $E[X_{(j)}^2] = 1, E[X_{(i)}X_{(j)}] = 0$ if $i \neq j$. (If the density function of X is differentiable the functions can be defined by

$$x_{(j)} = \frac{1}{p_X(x)} \frac{d^j p_X(x)}{dx^j}.)$$

Then (Lancaster [38]) the bivariate joint density function of X_1 and X_2 can be written

$$p_{X_1,X_2}(x_1,x_2) = p_{X_1}(x_1)p_{X_2}(x_2)\left[\sum_{j_1=0}^{\infty}\sum_{j_2=0}^{\infty} \rho_{(j_1,j_2)}x_{1(j_1)}x_{2(j_2)}\right], \tag{9}$$

where $x_{t(0)} = 1$, $\rho_{(00)} = 1$, and

$$\rho_{(j_1,j_2)} = E[X_{1(j_1)}X_{2(j_2)}]$$

is called the *generalized correlation coefficient* of order (j_1,j_2) between X_1 and X_2, provided that

$$\phi^2 = E\left[\frac{p_{X_1,X_2}(X_1,X_2)}{p_{X_1}(X_1)p_{X_2}(X_2)}\right] - 1 = \sum_{j_1+j_2>0}\sum \rho^2_{(j_1,j_2)} \tag{10}$$

is finite. (The ϕ^2 was originally introduced by Pearson as a "contingency coefficient." See also Hirschfeld [27].)

It follows directly that a necessary and sufficient set of conditions for such a continuous bivariate distribution to be symmetrical (in the case defined at the end of Section 1) is that

(a) the marginal distributions be identical and
(b) $\rho_{(j_1,j_2)} = \rho_{(j_2,j_1)}$ for all j_1,j_2.

Formula (9) can be extended in a rational fashion to the joint distribution of m random variables $X_1,\ldots,X_m$. The conditions for symmetry of an m-variate distribution with finite

$$\phi^2 = E\left[p_{\mathbf{X}}(\mathbf{X})\left\{\prod_{j=1}^{m} p_{X_j}(X_j)\right\}^{-1}\right] - 1 \tag{11}$$

are that $\rho_{(j_1,\ldots,j_m)}$ shall be unchanged for every permutation of $j_1,\ldots,j_m$, for any given set of values $(j_1,\ldots,j_m)$ (Lancaster [39], see also Eagleson [16]).

Several of the expansions encountered in this volume will be recognized as being of the kind just described.

Jensen [28] has shown that if two random variables X_1 and X_2 have a joint distribution that can be expanded in an orthonormal series with all the (generalized correlation) coefficients positive *and with identical marginal distributions*, then

$$\Pr[(X_1 \text{ in } A) \cap (X_2 \text{ in } A)] \geq \Pr[X_1 \text{ in } A]\Pr[X_2 \text{ in } A]$$

for all sets A for which the probabilities exist. The italicized condition can probably be relaxed in many cases.

Griffiths [24] has shown that under fairly broad conditions any sequence of positive numbers $\rho_1,\rho_2,\ldots$, with $\sum_{j=1}^{\infty} \rho_j^2$ finite can be canonical correlations of a symmetric bivariate distribution with ϕ^2 finite.

Mihaĭla [48] has given explicit formulas for Gram-Charlier expansions of trivariate density functions. In terms of standardized variables we can write

$$p(x_1,x_2,x_3) = \left[\sum_{j_1=0}^{\infty}\sum_{j_2=0}^{\infty}\sum_{j_3=0}^{\infty} C_{j_1,j_2,j_3} \frac{\partial^{j_1+j_2+j_3}}{\partial x_1^{j_1}\,\partial x_2^{j_2}\,\partial x_3^{j_3}}\right] Z_3(\mathbf{x};\mathbf{O};\mathbf{R}), \tag{12}$$

where $Z_3(\mathbf{x};\mathbf{O},\mathbf{R})$ is a standardized trivariate normal density function with correlation matrix $\mathbf{R}$ [Chapter 36, equation (3)]. The expansion can be expressed in terms of trivariate Hermite polynomials:

$$\frac{\partial^{j_1+j_2+j_3}}{\partial x_1^{j_1}\,\partial x_2^{j_2}\,\partial x_3^{j_3}} Z_3(\mathbf{x};\mathbf{O};\mathbf{R}) = (-1)^{j_1+j_2+j_3} H_{j_1,j_2,j_3}(\mathbf{x}) Z_3(\mathbf{x};\mathbf{O};\mathbf{R}). \tag{13}$$

These polynomials have coefficients that depend on the correlation matrix **R**. They are most conveniently expressed in terms of the elements of the inverse matrix $\mathbf{A} = \mathbf{R}^{-1}$. We have (up to the 4th order)

$$H_{000} = 1, \quad H_{100} = x_1, \quad H_{200} = x_1^2 - a_{11}, \quad H_{110} = x_1x_2 - a_{12},$$

$$H_{300} = x_1^3 - 3a_{11}x_1, \quad H_{210} = x_1^2x_2 - 2a_{12}x_1 - a_{11}x_2,$$

$$H_{111} = x_1x_2x_3 - a_{23}x_1 - a_{13}x_2 - a_{12}x_3,$$

$$H_{400} = x_1^4 - 6a_{11}x_1^2 + 3a_{11}^2, \quad H_{310} = x_1^3x_2 - 3a_{12}x_1^2 - 3a_{11}x_1x_2 + 3a_{11}a_{12}$$

$$H_{200} = x_1^2x_2^2 - a_{22}x_1^2 - a_{11}x_2^2 - 4a_{12}x_1x_2 + a_{11}a_{22} + 2a_{12}^2,$$

$$H_{211} = x_1^2x_2x_3 - a_{23}x_1^2 - 2a_{13}x_1x_2 - 2a_{12}x_1x_3 - a_{11}x_2x_3 + a_{11}a_{23} + 2a_{12}a_{13},$$

Other expressions can be obtained by permutation of subscripts; e.g.,

$$H_{201} = x_1^2x_3 - 2a_{13}x_1 - a_{11}x_3.$$

The coefficients $C_{j_1j_2j_3}$ are given by the following formulas, in which $\mu_{r_1r_2r_3}$ denotes

$$E\left[\prod_{j=1}^{3} \{X_j - E[X_j]\}^{r_j}\right].$$

(If a nonstandardized distribution is being fitted then $\mu_{r_1r_2r_3}$ should be replaced by $\beta_{r_1r_2r_3} = \mu_{r_1r_2r_3}/\sigma_1^{r_1}\sigma_2^{r_2}\sigma_3^{r_3}$ in an obvious notation.) $C_{000} = 1$; $C_{100} = 0$; $C_{200} = 0$; $C_{110} = \mu_{110} - \rho_{12}$ (Note that $C_{110} = 0$ if we choose ρ_{12} equal to the actual correlation between X_1 and X_2.)

$$C_{300} = -\tfrac{1}{6}\mu_{300}; \quad C_{210} = -\tfrac{1}{2}\mu_{210}; \quad C_{111} = \mu_{111};$$

$$C_{400} = \tfrac{1}{24}(\mu_{400} - 3); \quad C_{310} = \tfrac{1}{6}(\mu_{310} - 3\mu_{110});$$

$$C_{220} = \tfrac{1}{4}(\mu_{220} - \mu_{200} - \mu_{020} - 4\rho_{12}\mu_{110} - 1 + 2\rho_{12}^2);$$

$$C_{211} = \tfrac{1}{2}(\mu_{211} - \rho_{23}\mu_{200} - 2\rho_{13}\mu_{110} - 2\rho_{12}\mu_{101} - \mu_{011} + \rho_{23} + \rho_{12}\rho_{13}).$$

As in the case of the H's, further values can be obtained by permutation of the subscripts.

Formulas for the bivariate case can be obtained from those for trivariate distributions in the following simple way. To obtain H_{r_1,r_2}, C_{r_1,r_2} take the formula for $H_{r_1,r_2,0}, C_{r_1,r_2,0}$ respectively and replace $\mu_{s_1,s_2,0}$ by μ_{s_1,s_2}. [Of course a_{ij} are now elements of a 2×2 matrix and, in fact $a_{11} = a_{22} = (1 - \rho_{12}^2)^{-1}$; $a_{12} = -\rho_{12}(1 - \rho_{12}^2)^{-1}$.]

As m increases the algebra rapidly becomes more complex, but the formulas are similar and the method of fitting remains the same. (See also Guldberg [25] and Meixner [46a].)

Sarmonov and Bratoeva [71] find conditions for

$$\frac{1}{2\pi} e^{-\frac{1}{2}(x^2+y^2)}\left[1 + \sum_{j=1}^{\infty} C_j H_j(x) H_j(y)\right]$$

to be nonnegative for all x and y.

They show that a necessary and sufficient condition is that $\{C_j\}$ be the moment sequence of some distribution with range contained in the interval $[-1,1]$. Presumably a similar result holds for expansions with n variables.

Chambers [10] has given an algorithm for the construction of Edgeworth-type expansions (see Chapter 12, Section 4) for a general m-variate distribution with joint characteristic function $\varphi(\mathbf{t})$ and cumulant generating function $K(\mathbf{t}) = \log \varphi(\mathbf{t})$.

Generalizing (7) we define the m-variate Hermite polynomial $H(\mathbf{x};\mathbf{r};\mathbf{A})$ by the equation

$$H(\mathbf{x};\mathbf{r};\mathbf{A})\exp(-\tfrac{1}{2}\mathbf{x}'\mathbf{A}\mathbf{x}) = (-1)^{\Sigma r_j} \frac{\partial^{\Sigma r_j}}{\partial x_1^{r_1} \cdots \partial x_m^{r_m}} \exp(-\tfrac{1}{2}\mathbf{x}'\mathbf{A}\mathbf{x}). \tag{14}$$

The rth Edgeworth approximation to the joint density function $p_{\mathbf{X}}(\mathbf{x})$ is constructed as follows.

(i) Calculate the polynomial in $\mathbf{t} = (t_1,\ldots,t_m)$: $Q^{(r)}(\mathbf{t})$ = terms of $K(\mathbf{t})$ from order 3 to order $(r + 2)$ inclusive.

(ii) Expand $\{\exp Q^{(r)}(\mathbf{t})\}$ to terms of order $n^{-(1/2)r}$ assuming that cumulants of order s are of order $n^{1-(1/2)s}$ for $s \geq 2$, and first-order cumulants are of order $n^{-1/2}$. We denote the resulting expansion by

$$P^{(r)}(\mathbf{t}) = 1 + \sum_{j=1}^{r} P_j(\mathbf{t})n^{-(1/2)j}.$$

(iii) Replace the product $\prod_{j=1}^{m} t_j^{r_j}$ in $P^{(r)}(\mathbf{t})$ by $(-1)^{\Sigma r_j}H(\mathbf{x};\mathbf{r};\mathbf{V}^{-1})$ for all $\mathbf{r}$. Denote the result by $R_{(\mathbf{r})}(\mathbf{x})$. Then the required approximation is $R_{(r)}(\mathbf{x})Z_m(\mathbf{x};\mathbf{O},\mathbf{V})$.

Chambers [10] gives some account of formal convergence of the series expansion (though for practical purposes, when the expansion is fitted, rather few terms are used, and this question is of little importance). Bikelis [4] has studied the problem in some detail and has obtained the following result. If $X_1,\ldots,X_m$ are standardized variables and $\mathbf{X}_j = (X_{1j},\ldots,X_{kj})$ $(j = 1,2,\ldots,n)$ are independent vectors each having the same distribution as $\mathbf{X} = (X_1,\ldots,X_m)$, then the joint characteristic function of $\mathbf{S}_n = n^{-1/2} \sum_{j=1}^{n} \mathbf{X}_j$, which is $\{E[\exp(i\mathbf{t}'\mathbf{X}/\sqrt{n})]\}^n$, can be expressed in the form

$$e^{-(1/2)Q(\mathbf{t})}\left[1 + \sum_{j=1}^{s-3} P_j(i\mathbf{t})n^{-(1/2)j}\right] + R_s, \tag{15}$$

where the P_j's are certain polynomials, $Q(\mathbf{t})$ is a positive definite quadratic form in $\mathbf{t}$, and

$$|R_s| \leq (2/0.99)^{s-1}n^{-\frac{1}{2}(s+1)}E[|\mathbf{t}'\mathbf{X}|^s]e^{-(1/4)Q(t)}$$

provided that

$$\frac{E[|\mathbf{t}'\mathbf{X}|^s]}{Q(\mathbf{t})} \leq \left(\frac{\sqrt{n}}{8}\right)^{s-2} \tag{16}$$

(s can be chosen arbitrarily, but X must possess finite moments of order s. The condition (16) is a limitation on values of t.)

In [5–7] Bikelis uses (15) to obtain expansions for the difference between the densities, and between the cumulative distribution functions, for $\mathbf{S}_n$ and a multinormal distribution. In [6] he shows that if the joint density function of $\mathbf{X}$ has an upper bound C, and the expected value vector is $\mathbf{O}$, then the modulus of the characteristic function $|E[\exp(i\mathbf{t}'\mathbf{X})]|$ cannot exceed

$$\exp\left[-\frac{\pi^2}{27C^2}\cdot\frac{\{2^{m-1}(m-1)!\}^2\mathbf{t}'\mathbf{V}\mathbf{t}}{(8\pi)^m m^{m-1}\,|\mathbf{V}|\,\{2\pi+\sqrt{m}\sqrt{\mathbf{t}'\mathbf{V}\mathbf{t}}\}^2}\right].$$

He uses this expression to obtain another upper bound for $|R_s|$. In [7] Bikelis shows that for sums of n independent and identically distributed random vectors, with finite third moments, the error of approximation to the probability integral using a transformation truncated at $s = 4$ is $o(n^{-1/2})$ uniformly in all m variables. (See also Bikelis and Mogyoródi [8].)

This result is evidently a multivariate relationship analogous to the univariate central limit theorems. (Note that the characteristic function of a standardized multinormal distribution is of the form $e^{-(1/2)Q(t)}$, as will be seen in Chapter 35, Section 2.)

Among further work on multivariate central limit theorems, we take note of a paper by Sazonov [72]. He shows that provided all moments of the third order of $\mathbf{X}$ exist, the difference between the probability that $\mathbf{S}_n$ falls in a region E and the integral over E of the standardized multinormal density function [equation (13), Chapter 35] with the same correlation matrix as each $\mathbf{X}_j$ is less than

$$n^{-1/2}C(m,r)\sup_{\mathbf{l}\neq\mathbf{0}}\sqrt{\beta_1}\,(|\mathbf{l}'\mathbf{X}|),$$

where $C(m,r)$ is a constant that may depend on m and r, and E is the intersection of r sets defined by $\mathbf{a}_h'\mathbf{X} \gtrless \alpha_h$ $(h = 1,2,\ldots,r)$. Sazonov suggests that $C(m,r)$ may be replaced by a constant depending only on m. Paulauskas [55] has generalized these results to a wider class of sets E.

Zolotarev [84] has shown that, if $\mathbf{X}_1,\mathbf{X}_2,\ldots$ are a sequence of independent identically distributed m-dimensional vectors, composed of correlated elements each having zero mean and unit variance, then provided the vector lengths $|X_j|$ have finite fourth moments,

$$(17)\quad \lim_{n\to\infty} \sqrt{n} \sup_A \Big[\Pr[n^{-1/2} \sum \mathbf{X}_j \in A] - (2\pi)^{-(1/2)m} \times \int \cdots \int_A \exp\Big(-\tfrac{1}{2}\sum_{j=1}^{m} x_j^2\Big)\, dx_1 \ldots dx_m \Big] \le \frac{1}{6\sqrt{2\pi}} [1 + 2(1+\theta)e^{-3/2}]\nu_3,$$

where $\nu_3 = E[|\mathbf{X}_j|^3]$ and $\theta = 0$ or 1 according as the regions A are restricted to being simply connected or not. Note that the upper bound in (17) does not depend on m. (Of course, the region A is supposed to be such that the integral in (17) exists.)

Dunnage [14] has shown that if (X_{ij}, X_{2j}) each have expected value vector (0,0) and are mutually independent ($j = 1,2,\ldots,n$) and $S_{in} = \sum_{j=1}^{n} X_{ij}$ ($i = 1,2$) then, for all s_1, s_2,

$$(18)\quad \left| F_{S_{1n},S_{2n}}(s_1,s_2) - \Phi\left(\frac{s_1}{\sigma_1}, \frac{s_2}{\sigma_2}; \rho_n\right)\right| \le K \cdot \frac{\nu_3^{1/3}}{\min(\sigma_1,\sigma_2)} + \frac{n\nu^{1/2}}{k^{3/2}} + \frac{n\nu \log n}{k^{3/2}},$$

where K is an absolute constant and

$$\sigma_i^2 = \sum_{j=1}^{n} \text{var}(X_{ij}) = \text{var}(S_{in}); \qquad \rho_n = \text{correlation between } S_{1n} \text{ and } S_{2n};$$

$$\nu_3 = \max_{i,j} E[|X_{ij}|^3];$$

$$\nu = n^{-1}\sum_{j=1}^{n} \max\{E[|X_{1j}|^3], E[|X_{2j}|^3]\};$$

and

$$k = \frac{1}{\sqrt{2}}[\sigma_1^2 + \sigma_2^2 - \{(\sigma_1^2 - \sigma_2^2)^2 + 4\rho_n^2\sigma_1^2\sigma_2^2\}^{1/2}].$$

Note that this result still holds, even if some of the correlations between X_{ij}, X_{2j} are numerically equal to 1. Note also that k lies between $\frac{1}{2}(1 - \rho_n^2)\min(\sigma_1^2, \sigma_2^2)$ and $(1 - \rho_n^2)\min(\sigma_1^2, \sigma_2^2)$.

In [15], Dunnage has shown that the right-hand side of (18) may be replaced by

$$\frac{n\nu}{\{\min(\sigma_1,\sigma_2)\}^3(1-\rho_n^2)^{3/2}}\left[24 + \tfrac{4}{5}\log\left\{\frac{(\min(\sigma_1,\sigma_2))^3(1-\rho_n^2)^{1/2}}{n\nu}\right\}\right] + \max\left[\frac{2\nu_3^{1/2}}{\min(\sigma_1,\sigma_2)}, \frac{48n\nu}{\{\min(\sigma_1,\sigma_2)\}^3}(1-\rho_n^2)\right]$$

The concept of a *stable* distribution (Chapter 19, Section 8) can be directly generalized to sets of m variables. If $\mathbf{X}_1,\mathbf{X}_2$, and $\mathbf{X}$ have the same joint distribution, with $\mathbf{X}_1,\mathbf{X}_2$ independent, and for any nonsingular $\mathbf{A}_1,\mathbf{A}_2,\mathbf{B}_1(m \times m)$ and $\mathbf{B}_2(m \times m)$ it is possible to find $\mathbf{A},\mathbf{B}$ such that $\mathbf{B}(\mathbf{X} - \mathbf{A})$ has the same distribution as $\mathbf{B}_1(\mathbf{X}_1 - \mathbf{A}_1) + \mathbf{B}_2(\mathbf{X}_2 - \mathbf{A}_2)$, then the common joint distribution is said to be *stable*. The general forms of such distributions are discussed by Kalinauskaitė [32, 33]. In [33] he discusses the *symmetrical stable distributions*, which have characteristic functions $\exp\left[-\left(\sum_{j=1}^{n} t_j^2\right)^{\alpha/2}\right]$ $(0 < \alpha \le 2)$. The case $\alpha = 2$ gives a multinormal distribution; for $0 < \alpha < 2$ the joint density function is

$$\alpha^{-1}(2\pi)^{-\frac{1}{2}m} \sum_{j=0}^{\infty} (-\tfrac{1}{4})^j \frac{\Gamma((2j + s)\alpha^{-1})}{\Gamma(j + 1)\Gamma(j + \frac{1}{2}s)} \left(\sum_{j=1}^{m} x_j\right)^j .$$

5. Translation Systems

We have already described (in Chapter 12, Section 4.3) systems of distributions constructed by supposing certain (fairly simple) functions of variables to be normally distributed. It is natural to consider what forms of joint distribution one can construct in this way. We first suppose that it is possible to normalize the marginal distributions by simple univariate transformations, and then consider how to construct joint distributions with such marginal distributions.

If each of a set of m variables is normally distributed it is not necessary that their joint distribution should be multinormal. (An example is given in Chapter 42, Section 7.) However, it is possible for this to be so, and some systems of distributions have been constructed in this way.

It is interesting to note that if $Y_1 = f_1(X_1)$ and $Y_2 = f_2(X_2)$ do have a joint bivariate normal distribution, then the absolute value of the correlation between Y_1 and Y_2 is the greatest that can be attained for any choice of functions $g_1(X_1),g_2(X_2)$ (Maung [46], Lancaster [36]).

Johnson [29] has studied bivariate distributions, S_{IJ}, in which one variable X_1 has an S_I distribution and the other, X_2, an S_J distribution, where I,J can take the values B, U, L, and N. (S_B,S_U are defined in Chapter 12, Section 4.3; S_L means "log-normal"; S_N means "normal"). Thus the variables

$$Z_1 = \gamma_1 + \delta_1 f_I((X_1 - \xi_1)/\lambda_1)$$
$$Z_2 = \gamma_2 + \delta_2 f_J((X_2 - \xi_2)/\lambda_2)$$

(where $f_B(y) = \log\{y/(1 - y)\}$, $f_U(y) = \sinh^{-1} y$, $f_L(y) = \log y$, and $f_N(y) = y$) are standardized (unit normal) variables, with a joint bivariate normal distribution with correlation coefficient ρ. (We also take $\delta_1 > 0$, $\delta_2 > 0$ by convention.)

The joint distribution of X_1 and X_2, so defined, has nine parameters $\gamma_1, \gamma_2, \delta_1, \delta_2, \xi_1, \xi_2, \lambda_1, \lambda_2$, and ρ. The *shape* of the distribution depends only on the five parameters γ_1, γ_2, δ_1, δ_2, and ρ. The *standard form* of the distribution is obtained by taking $\xi_1 = \xi_2 = 0$; $\lambda_1 = \lambda_2 = 1$. This form, which is convenient for algebraic treatment, will be used in the discussion that follows.

The random variable

$$T = (1 - \rho^2)^{-1}(Z_1^2 - 2\rho Z_1 Z_2 + Z_2^2)$$

has a χ^2 distribution with 2 degrees of freedom. It is thus quite easy to construct regions in the X_1,X_2 plane containing specified proportions (α) of the distribution, which have boundaries with equations

$$(19) \quad (1 - \rho^2)^{-1}[\{\gamma_1 + \delta_1 f_I(X_1)\}^2 - 2\rho\{\gamma_1 + \delta_1 f_J(X_1)\}\{\gamma_2 + \delta_2 f_J(X_2)\} + \{\gamma_2 + \delta_2 f_J(X_2)\}^2] = \chi^2_{2,\alpha} = -2 \log(1 - \alpha).$$

It should be realized, however, that these boundaries are not (except in the case of the bivariate normal distribution) contours on which the probability density function is constant.

The conditional distribution of Z_2, given Z_1, is normal with expected value ρZ_1 and standard deviation $\sqrt{1 - \rho^2}$ (see Chapter 36, Section 3). This is thus the conditional distribution of $\gamma_2 + \delta_2 f_J(X_2)$ given X_1. The conditional distribution of

$$(1 - \rho^2)^{-1/2}[\gamma_2 - \rho(\gamma_1 + \delta_1 f_I(X_1)) + \delta_2 f_J(X_2)]$$

is therefore standard normal. This means that the conditional (array) distribution of X_2, given X_1, is of the same system (S_J) as X_2, but with γ_2,δ_2 replaced by $(1 - \rho^2)^{-1/2}[\gamma_2 - \rho\{\gamma_1 + \delta_1 f_I(X_1)\}]$, $(1 - \rho^2)^{-1/2}\,\delta_2$ respectively. All these array distributions have the same δ-parameter, but (if $\rho \neq 0$) the γ parameter varies from $-\infty$ to $+\infty$. The shape (and variance) of the array distributions of X_2 therefore change with X_1. In particular when the sign of the skewness depends on the γ-parameter (e.g., when $I \equiv U$ or $I \equiv B$) there will be a change in sign of skewness of the array distributions. This feature is, in fact, observed in empirical joint distributions.

A further consequence is that when $J \equiv B$ there may be a range of values of X_1 for which the array distribution of X_2 is bimodal but outside of which it is unimodal.

It is an easy matter to calculate any required percentage points of the array distributions from the formula:

$$(20) \qquad f_J(X_{2,\alpha}) = [U_\alpha\sqrt{1 - \rho^2} - \gamma_2 + \rho\{\gamma_1 + \delta_1 f_I(X_1)\}]\,\delta_2^{-1}.$$

In particular we have the median regression

$$M(X_2 \mid X_1) = f_J^{-1}[(\rho\gamma_1 - \gamma_2)\delta_2^{-1} + (\rho\delta_1/\delta_2)f_I(X_1)]. \tag{21}$$

This depends on I,J and the two parameters $\gamma = (\rho\gamma_1 - \gamma_2)\,\delta_2^{-1}$; $\phi = \rho\delta_1/\delta_2$.

Taking all possible combinations ($I,J \equiv N,L,B,U$) we have 16 possible median regression functions which are set out in Table 2 (taken from [29]).

TABLE 2

Median Regressions for S_{IJ} Distributions

Distribution of X_2	Distribution of X_1	Median of X_2 when $X_1 = x_1$
S_N	S_N	$\log\theta + \phi x_1$
S_N	S_L	$\log\theta + \phi\log x_1$
S_L	S_N	$\theta e^{\phi x_1}$
S_N	S_B	$\log\theta + \phi\log\{x_1/(1 - x_1)\}$
S_B	S_N	$[1 + \theta^{-1}e^{-\phi x_1}]^{-1}$
S_N	S_U	$\log\theta + \phi\log[x_1 + \sqrt{(x_1^2 + 1)}]$
S_U	S_N	$\frac{1}{2}[\theta e^{\phi x_1} - \theta^{-1}e^{-\phi x_1}]$
S_L	S_L	θx_1^ϕ
S_L	S_B	$\theta[x_1/(1 - x_1)]^\phi$
S_B	S_L	$[1 + \theta^{-1}x_1^{-\phi}]^{-1}$
S_L	S_U	$\theta[x_1 + \sqrt{(x_1^2 + 1)}]^\phi$
S_U	S_L	$\frac{1}{2}[\theta x_1^\phi - \theta^{-1}x_1^{-\phi}]$
S_B	S_B	$\theta x_1^\phi[(1 - x_1)^\phi + \theta x_1^\phi]^{-1}$
S_B	S_U	$[1 + \theta^{-1}\{\sqrt{(x_1^2 + 1)} - x_1\}^\phi]^{-1}$
S_U	S_B	$\frac{1}{2}[\theta x_1^{2\phi} - \theta^{-1}(1 - x_1)^{2\phi}]x_1^{-\phi}(1 - x_1)^{-\phi}$
S_U	S_U	$\frac{1}{2}[\theta\{x_1 + \sqrt{(x_1^2 + 1)}\}^\phi - \theta^{-1}\{\sqrt{(x_1^2 + 1)} - x_1\}^\phi]$

(Note: $\theta = \exp[(\rho\gamma_1 - \gamma_2)/\delta_2]$; $\phi = \rho\delta_1/\delta_2$.)

The graphical forms of these regressions are shown in Figure 1 (also taken from [29]). The examples in these diagrams should suffice to show the effect of reversing the sign of ϕ.

The system S_{NL} ("normal lognormal") has been discussed in some detail by Crofts [12], who gives formulas for maximum likelihood estimators of the parameters. Crofts also considers the more general estimation when X_1 is normal, the conditional distribution of X_2 given X_1 is (three-parameter) lognormal, and the regression of X_2 on X_1 has a specified form.

In the bivariate lognormal distribution (S_{LL}) it is possible to obtain reasonably simple expressions for the ordinary regression function. Assuming that $\log X_1, \log X_2$ have a joint bivariate normal distribution with expected

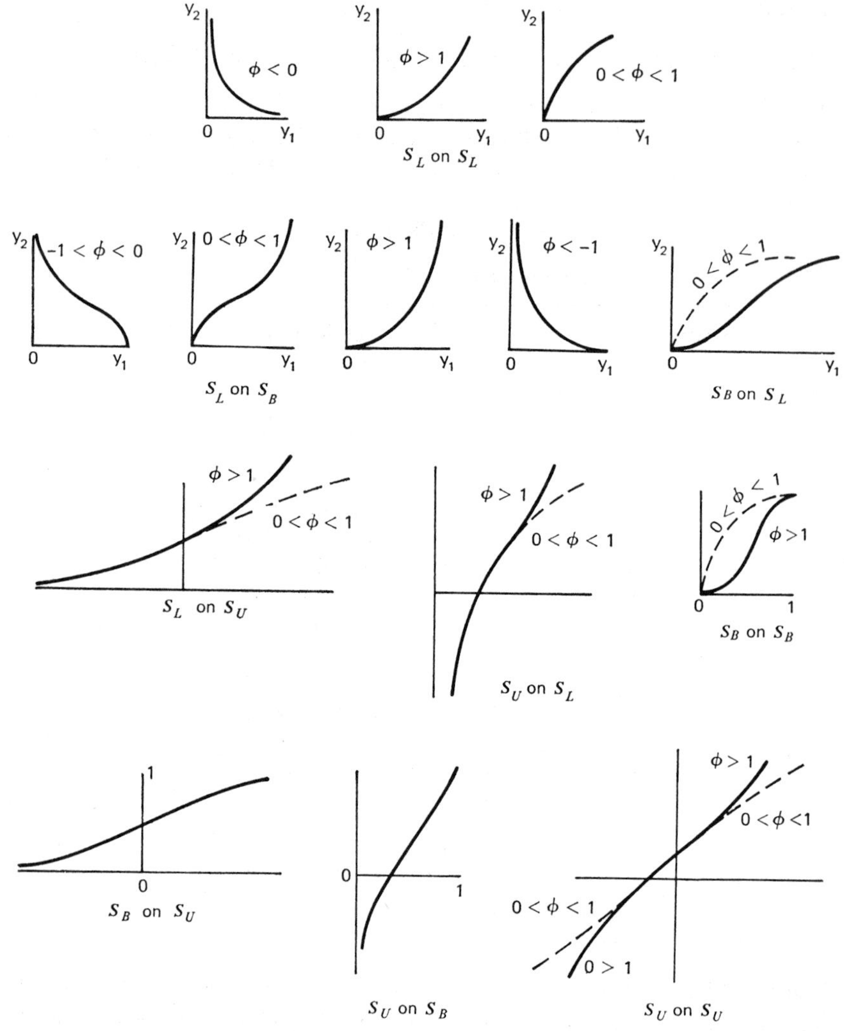

FIGURE 1

values (ζ_1, ζ_2), variances (σ_1^2, σ_2^2), and correlation ρ, the conditional distribution of $\log X_2$ given X_1 is normal with expected value

$$\zeta_2 + (\rho\sigma_2/\sigma_1)(\log X_1 - \zeta_1) = \zeta_2(X_1)$$

and variance $(1 - \rho^2)\sigma_2^2 = \sigma_2'^2$. The conditional distribution of X_2, given X_1, is therefore lognormal with parameters $\zeta_2(X_1), \sigma_2'$. It follows that the regression of X_2 on X_1 is [from (6.1) of Chapter 14]

(22.1) $$E[X_2 \mid X_1] = \exp(\zeta_2(X_1) + \tfrac{1}{2}\sigma_2'^2)$$
$$= X_1^{\rho\sigma_2/\sigma_1} \exp[\tfrac{1}{2}(1 - \rho^2)\sigma_2^2 + \zeta_2 - \rho\sigma_2\zeta_1/\sigma_1].$$

The array variance is

(22.2) $$\operatorname{var}(X_2 \mid X_1) = \omega'(\omega' - 1)X_1^{2\rho\sigma_2/\sigma_1} \exp[2(\zeta_2 - \rho\sigma_2\zeta_1/\sigma_1)],$$

where $\omega' = \exp[(1 - \rho^2)\sigma_2^2]$.

Mostafa and Mahmoud [50] have constructed an unbiased estimator of the function (22.1), based on a random sample giving n pairs of observed values (X_{1j}, X_{2j}) $(j = 1, 2, \ldots, n)$ of X_1, X_2. Since $(Z_{1j}, Z_{2j}) = (\log X_{1j}, \log X_{2j})$ have a joint bivariate normal distribution, we can obtain maximum likelihood estimators of ζ_1, ζ_2, σ_1, σ_2, and ρ using the standard formulas given in Chapter 36, Section 6. Mostafa and Mahmoud [50] show that

$$E[\exp\{\hat{\zeta}_2 + (\hat{\rho}\hat{\sigma}_2/\hat{\sigma}_1)(\log X_1 - \hat{\zeta}_1)\}] = \exp[\zeta_2(X_1) + \tfrac{1}{2}\sigma_2'^2 n^{-1}(1 + nK)],$$

where

$$K = (\log X_1 - \zeta_1)^2 \left[\sum_{j=1}^{n} (\log X_{1j} - \zeta_1)^2\right]^{-1}.$$

They then seek to find a function $g(\hat{\sigma}_2'^2)$ of the residual mean square

$$\hat{\sigma}_2'^2 = (n - 2)^{-1} \sum_{j=1}^{n} \{\log X_{2j} - \hat{\zeta}_2 - (\hat{\rho}\hat{\sigma}_2/\hat{\sigma}_1)(\log X_{1j} - \hat{\zeta}_1)\}^2,$$

which shall have expected value $\exp[\tfrac{1}{2}\{1 - n^{-1}(1 + nK)\}\sigma_2'^2]$. Then, since $\hat{\sigma}_2'^2$ and $[\hat{\zeta}_2 + (\hat{\rho}\hat{\sigma}_2/\hat{\sigma}_1)(\log X_1 - \hat{\zeta}_1)]$ are independent, the product

(23) $$g(\hat{\sigma}_2'^2)\exp[\hat{\zeta}_2 + (\hat{\rho}\hat{\sigma}_2/\hat{\sigma}_1)(\log X_1 - \hat{\zeta}_1)]$$

will be an unbiased estimator of $E[X_2 \mid X_1]$. They find

(24) $$g(\hat{\sigma}_2'^2) = \sum_{j=0}^{\infty} (\lambda^j/j!)\{(n - 2)\hat{\sigma}_2'^2\}^j \Gamma(\tfrac{1}{2}n - 1)\{\Gamma(\tfrac{1}{2}n + j - 1)\}^{-1},$$

where $\lambda = \tfrac{1}{2}[1 - n^{-1}(1 + nK)]$. For practical calculation they suggest the approximation

(24)′ $$g(\hat{\sigma}_2'^2) = [1 - n^{-1}\{\hat{\sigma}_2'^2 + (1 - K)^2\hat{\sigma}_2'^4\}]\exp[(1 - K)\hat{\sigma}_2'^2].$$

The variance of the estimator is approximately

(25) $$[E[X_2 \mid X_1]]^2[\{1 + n^{-1}\sigma_2'^2 + \tfrac{1}{2}n^{-1}(1 - K)^2\sigma_2'^4\}\exp(K\sigma_2'^2) - 1].$$

Mostafa and Mahmoud [50] also give formulas for estimators of the median regression and of the *modal regression:*

(26) $$\operatorname{Mode}[X_2 \mid X_1] = \exp[\zeta_2 + (\rho\sigma_2/\sigma_1)(\log X_1 - \zeta_1) - \sigma_2^2(1 - \rho^2)].$$

To form the *m-variate lognormal distribution* we suppose the variables $Z_j = \log X_j$ $(j = 1,\ldots,m)$ to have a joint multinormal distribution with expected value vector $\boldsymbol{\zeta} = (\zeta_1,\ldots,\zeta_m)$ and variance-covariance matrix V.

By a similar analysis to that used in the bivariate case, it can be seen that the conditional distribution of X_1, given $X_2,\ldots,X_n$, is lognormal. The moments and product moments of the X's are derived straightforwardly from the moment-generating function of the Z's, since

$$\mu_{r_1,r_2,\ldots,r_m}(\mathbf{X}) = E\left[\prod_{j=1}^{m} X_j^{r_j}\right] \tag{27}$$

$$= E[\exp(\mathbf{r}'\mathbf{Z})]$$

$$= \exp(\mathbf{r}'\boldsymbol{\zeta} + \tfrac{1}{2}\mathbf{r}'\mathbf{V}\mathbf{r}).$$

(from equation (4) of Chapter 35).

Putting $r_j = r_k = 1$ and all the other r's equal to zero, we find

$$\text{cov}(X_j,X_k) = \{\exp(\rho_{jk}\sigma_j\sigma_k) - 1\}\exp[\zeta_j + \zeta_k + \tfrac{1}{2}(\sigma_j^2 + \sigma_k^2)], \tag{28.1}$$

[where $\rho_{jk} = \text{corr}(Z_j,Z_k)$], from which

$$\text{corr}(X_j,X_k) = \{\exp(\rho_{jk}\sigma_j\sigma_k) - 1\}[\{\exp(\sigma_j^2) - 1\}\{\exp(\sigma_k^2) - 1\}]^{-1/2} \tag{28.2}$$

(Jones and Miller [30]).

R. L. Obenchain (personal communication, Bell Telephone Laboratories) has suggested another multivariate extension of the S_B distributions, which might be appropriate when the range of variation is restricted to a simplex (e.g., $0 \leq \sum_{j=1}^{m} X_j \leq 1; X_j > 0,\ j = 1,\ldots,m$). He considers the joint distribution of

$$X_j = e^{Y_j}\left[\sum_{i=1}^{m+1} e^{Y_i}\right]^{-1} \qquad (j = 1,\ldots,m)$$

when $Y_1, Y_2,\ldots, Y_{m+1}$ have a multivariate normal distribution. By putting $Y_i^* = Y_i - Y_{m+1}$, we have

$$X_j = e^{Y_j^*}\left[1 + \sum_{i=1}^{m} e^{Y_i^*}\right]^{-1} \qquad (j = 1,\ldots,m) \tag{29}$$

with $Y_1^*,\ldots,Y_m^*$ jointly multinormally distributed. In the case $m = 1$, (29) gives an S_B distribution. For $m \geq 2$, the marginal distributions are not, in general, S_B.

Obenchain has made some detailed investigations of the bivariate ($m = 2$) case and has developed methods of fitting the distributions to data (i.e., estimating parameters).

6. Multivariate Exponential-Type Distributions

Bildikar and Patil [9] define an m-variate *exponential-type* distribution, in a general way, as a distribution with joint likelihood function of form

$$L_{\mathbf{X}}(\mathbf{x}) = h(\mathbf{x})\exp[\mathbf{x}'\mathbf{t} - q(\boldsymbol{\theta})], \tag{30}$$

where $\mathbf{X}' = (X_1,\ldots,X_m)$ represents random variables and $\boldsymbol{\theta}' = (\theta_1,\ldots,\theta_m)$ represents parameters. We are concerned here with continuous multivariate distributions, and so regard $L_{\mathbf{X}}(\mathbf{x})$ as a density function. (If the X's are discrete, we regard $L_{\mathbf{X}}(\mathbf{x})$ as the probability of the event $\bigcap_{j=1}^{m} (X_j = x_j)$.) The results of this section apply also to distributions that include multinomial, negative binomial, and multivariate series distributions, among others.

The following results are obtained in [9]. The moment-generating function of $X_1,\ldots,X_m$ is

$$\begin{aligned} E[\exp(\mathbf{t}'\mathbf{X})] &= e^{-q(\boldsymbol{\theta})}\int_{-\infty}^{+\infty}\cdots\int_{-\infty}^{\infty} h(\mathbf{x})\exp[\mathbf{x}'(\boldsymbol{\theta}+\mathbf{t})]\,d\mathbf{x} \\ &= \exp[q(\boldsymbol{\theta}+\mathbf{t}) - q(\boldsymbol{\theta})] \end{aligned} \tag{31.1}$$

(since $\int_{-\infty}^{\infty}\cdots\int_{-\infty}^{\infty} L_{\mathbf{X}}(\mathbf{x},\boldsymbol{\theta})\,d\mathbf{x} = 1$ for all $\boldsymbol{\theta}$). The cumulant generating function is

$$\Psi(\mathbf{t}) = q(\boldsymbol{\theta}+\mathbf{t}) - q(\boldsymbol{\theta}). \tag{31.2}$$

From (31.2) one can deduce the recurrence relation

$$\kappa_{r_1,\ldots,r_{j-1},r_j+1,r_{j+1},\ldots,r_m} = \partial\kappa_{r_1,\ldots,r_m}/\partial\theta_j \tag{32}$$

[cf. equation (21) of Chapter 2].

Taking $m = 2$, we see that if κ_{11} is zero (i.e., X_1 and X_2 are uncorrelated), then so are κ_{21} and κ_{12}, hence also κ_{r_1,r_2} for any $r_1 \geq 1$, $r_2 \geq 1$. In fact X_1 and X_2 are independent. This property can be extended: if $X_1,\ldots,X_m$ have a joint exponential-type distribution, they form a mutually independent set if and only if they are pairwise independent.

If

$$p(x_1,\ldots,x_m) = h(x_1,\ldots,x_m)\exp\left[\sum_{j=1}^{m}\theta_j x_j - q(\theta_1,\ldots,\theta_m)\right], \tag{33}$$

then (for $s < m$) the joint density function of $X_1,\ldots,X_s$ has the form

$$p(x_1,\ldots,x_s) = h_1(x_1,\ldots,x_s)\exp\left[\sum_{j=1}^{s}\theta_j x_j - q_1(\theta_1,\ldots,\theta_s)\right], \tag{34}$$

where

$$h_1(x_1,\ldots,x_s) = \int_{-\infty}^{\infty}\cdots\int_{-\infty}^{\infty} h(x_1,\ldots,x_m)\exp\left(\sum_{j=s+1}^{m}\theta_j x_j\right) dx_{s+1}\cdots dx_m$$

and $q_1(\theta_1,\ldots,\theta_s)$ is simply $q(\theta_1,\ldots,\theta_m)$ regarded as a function of $\theta_1,\ldots,\theta_s$ (i.e., $\theta_{s+1},\ldots,\theta_m$ regarded as pure constants, rather than parameters).

Comparison of (33) with (34) shows that the joint distribution of $X_1,\ldots,X_s$ (hence of any subset of $X_1,\ldots,X_m$) is of the exponential type.

The conditional joint density of $X_{s+1},\ldots,X_m$, given $X_1 = x_1, X_2 = x_2, \ldots, X_s = x_s$, is of form

$$h_s(x_{s+1},\ldots,x_m)\exp\left(\sum_{j=s+1}^{m}\theta_j x_j\right), \tag{35}$$

since $q_1(\boldsymbol{\theta}) \equiv q(\boldsymbol{\theta})$. (The function $h_s(\cdot)$ depends also on $x_1,\ldots,x_s$, but the θ's do not.) This is also of exponential type, but is restricted by the requirement that $q(\theta) \equiv 0$.

Seshadri and Patil [73] have shown that if $p_{X_1}(x_1)$ and $p_{X_1|X_2}(x_1 \mid x_2)$ are given, a sufficient condition for $p_{X_2}(x_2)$ to be unique is that the conditional density function $p_{X_1|X_2}(x_1 \mid x_2)$ is of (univariate) exponential form. Roux [66] has extended this result to sets of $m(\geq 2)$ variables. He has shown that given $p_{X_1,\ldots,X_{m-1}}(x_1,\ldots,x_{m-1})$ and $p_{X_1,\ldots,X_{m-1}|X_m}(x_1,\ldots,x_{m-1} \mid x_m)$ a sufficient condition for $p_{X_m}(x_m)$ [hence $p_{X_1\cdots X_m}(x_1,\ldots,x_m)$] to be unique is that the conditional density function $p_{X_1,\ldots,X_{m-1}|X_m}(x_1,\ldots,x_{m-1} \mid x_m)$ is of exponential form.

In view of the occurrence of a linear function of $\mathbf{x}$ in the exponent of (35), such distributions may be called *linear exponential-type* distributions (cf. Wani [78]) to distinguish them from the more general *quadratic exponential-type* distributions (Day [13]) for which

$$L_{\mathbf{X}}(\mathbf{x}) = h(\mathbf{x})\exp[-(\mathbf{x} - \boldsymbol{\xi})'\mathbf{A}(\mathbf{x} - \boldsymbol{\xi}) - q(\boldsymbol{\theta})] \tag{36}$$

where $\mathbf{A}$ is positive definite and $\mathbf{A}$, $\boldsymbol{\xi}$, and $\boldsymbol{\theta}$ are sets of parameters.

Day [13] has pointed out that if the optimal discriminant between two members of a parametric family is a linear (or quadratic) function of the observed values, then the family must be linear (or quadratic) exponential-type, respectively.

7. Other Methods of Constructing Multivariate Distributions

Fréchet [21] noted that since

$$\Pr[(X_1 \leq x_1) \cap (X_2 \leq x_2)] \leq \min[\Pr[X_1 \leq x_1], \Pr[X_2 \leq x_2]],$$

the relationship

$$(37.1)\qquad F_{X_1,X_2}(x_1,x_2) \le \min[F_{X_1}(x_1),F_{X_2}(x_2)]$$

must hold for all pairs of random variables, and all x_1,x_2.

In a similar way, since

$$\Pr[(X_1 > x_1) \cup (X_2 > x_2)] \le \Pr[X_1 > x_1] + \Pr[X_2 > x_2],$$

it follows that

$$1 - F_{X_1,X_2}(x_1,x_2) \le \{1 - F_{X_1}(x_1)\} + \{1 - F_{X_2}(x_2)\},$$

that is

$$(37.2)\qquad F_{X_1,X_2}(x_1,x_2) \ge F_{X_1}(x_1) + F_{X_2}(x_2) - 1.$$

Fréchet [21] suggested that any system of bivariate distributions with specified marginal distributions $F_{X_1}(x_1),F_{X_2}(x_2)$ should include the limits in (37) as limiting cases. In particular he suggested the system

$$(38)\quad F_{X_1,X_2}(x_1,x_2) = \theta \max(F_{X_1}(x_1) + F_{X_2}(x_2) - 1,0) + (1 - \theta)\min(F_{X_1}(x_1),F_{X_2}(x_2)) \qquad (0 \le \theta \le 1).$$

This system does not, however, include the case when X_1 and X_2 are independent. A system that does include this case [but not the limits in (37)] is

$$(39)\quad F_{X_1,X_2}(x_1,x_2) = F_{X_1}(x_1)F_{X_2}(x_2)[1 + \alpha\{1 - F_{X_1}(x_1)\}\{1 - F_{X_2}(x_2)\}].$$

This was introduced by Morgenstern [49] and later extended by Farlie [20] to the form

$$(39)'\qquad F_{X_1,X_2}(x_1,x_2) = F_{X_1}(x_1)F_{X_2}(x_2)[1 + \alpha f_1(x_1)f_2(x_2)],$$

where $f_1(x_1),f_2(x_2)$ are more general than $1 - F_{X_1}(x_1),1 - F_{X_2}(x_2)$. Mardia [44] (see also Nataf [52]) has pointed out that there is a simple way of constructing systems that include the limits in (37) and also the case of independence. This is done by finding the transformations $Y_j = g_j(X_j)(j = 1,2)$ which make Y_1, Y_2 unit normal variables (as in Section 5) and then ascribing a joint bivariate normal distribution to Y_1 and Y_2. (If X_1 and X_2 are each continuous random variables, there is always such a pair of transformations, defined by $F_{X_j}(x_j) = \Phi(g(x_j))(j = 1,2)$.) It is not necessary that the transformation be to bivariate normality. Many other standard joint distributions may be used.

Plackett [59] has constructed another such system which has some intrinsic interest, though it is more complicated than Mardia's system. The joint cumulative distribution function $F_{X_1,X_2}(x_1,x_2)$ is required to satisfy the equation

$$\psi = \frac{F_{12}(1 - F_1 - F_2 + F_{12})}{(F_1 - F_{12})(F_2 - F_{12})} \quad \text{with} \quad \psi > 0. \tag{40}$$

[The F_{12}, F_1, and F_2 are used, for convenience, to represent $F_{X_1,X_2}(x_1,x_2)$, $F_{X_1}(x_1)$, and $F_{X_2}(x_2)$, respectively.] For different values of X_1 different members of the system are obtained. If $\psi = 1$ in (40), then $F_{12} = F_1F_2$ and so X_1 and X_2 are independent. If $\psi = 0$ then F_{12} equals the lower limit in (37.2); if $\psi = \infty$ then F_{12} equals the upper limit in (37.1). In general there is just one value of F_{12}, in the interval defined in (37), which satisfies (40). This can be seen noting that, for F_1,F_2 fixed, the right-hand side of (40) is an increasing function of F_{12}, increasing from 0 to ∞ as F_{12} increases from the lower to the upper limit of (37). Furthermore, as F_1 increases (F_2 fixed), F_{12} increases; also as F_2 increases (F_1 fixed) F_{12} increases. We note that when x_1,x_2 take their median values, so that $F_{X_1}(x_1) = F_{X_2}(x_2) = \frac{1}{2}$, then $F_{X_1,X_2}(x_1,x_2) = \frac{1}{2}\sqrt{\psi}\,(1 + \sqrt{\psi})^{-1}$.

The conditional cumulative distribution function of X_1, given $X_2 = x_2$, is

$$\Pr[X_1 \le x_1 \mid X_2 = x_2] = \frac{\psi F_{X_1}(x_1) - (\psi - 1)F_{X_1,X_2}(x_1,x_2)}{1 + (\psi - 1)[F_{X_1}(x_1) + F_{X_2}(x_2) - 2F_{X_1,X_2}(x_1,x_2)]}. \tag{41}$$

The median regression $X_{1,0.5}(x_2)$ of X_1 on X_2 is obtained by equating this to $\frac{1}{2}$, leading to

$$(\psi + 1)F_{X_1}(X_{1,0.5}(x_2)) = 1 + (\psi - 1)F_{X_2}(x_2).$$

Note that as $F_{X_2}(x_2) \to 0$ [so that $F_{X_1,X_2}(x_1,x_2) \to 0$ also],

$$\Pr[X_1 \le x_1 \mid X_2 = x_2] \to \frac{\psi F_{X_1}(x_1)}{1 + (\psi - 1)F_{X_1}(x_1)}, \tag{42.1}$$

i.e., there is a nondegenerate limiting conditional distribution. Similarly, as $F_{X_2}(x_2) \to 1$ [and so $F_{X_1,X_2}(x_1,x_2) \to F_{X_1}(x_1)$],

$$\Pr[X_1 \le x_1 \mid X_2 = x_2] \to F_{X_1}(x_1)\{1 + (\psi - 1)(1 - F_{X_1}(x_1))\}^{-1}. \tag{42.2}$$

If $F_{X_j}(x_j)$ be taken equal to $\Phi(x_j)$ for $j = 1,2$, we have a bivariate distribution with unit normal marginal distributions which differs from the standardized bivariate normal distribution. Plackett [59] provides numerical comparisons of the two distributions. He also discusses estimation of ψ. It is clear from (40) that, for any double dichotomy as in Figure 2, $\tilde{\psi} = ad/bc$ (where a, b, c, and d are the observed frequencies in the cells indicated) should give a good

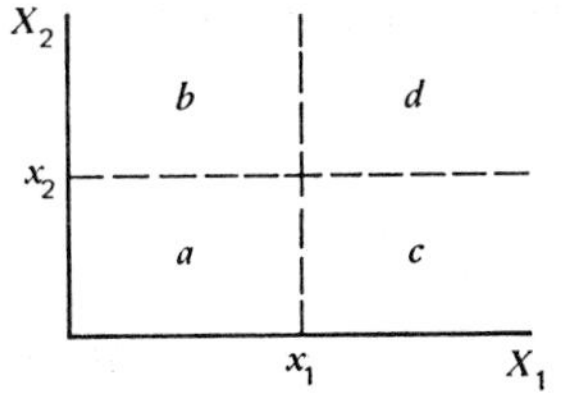

FIGURE 2

estimator of ψ. The variance of ψ may be estimated by

$$\tilde{\psi}^2(a^{-1} + b^{-1} + c^{-1} + d^{-1}).$$

Note that these formulas do not require a knowledge of $F_{X_1}(x_1), F_{X_2}(x_2)$. The functions can be estimated separately from the observed marginal distributions.

8. Multivariate Chebyshev-Type Inequalities

Inequalities satisfied by univariate distribution functions under fairly general conditions have been described in Chapter 33. We now describe some extensions of these inequalities to multivariate distributions. These are naturally more complicated, and often present more difficulties to intuitive comprehension than do the univariate inequalities. Whereas the univariate formulas are usually expressed in terms of moments of a single variable, it is only to be expected that to obtain good inequalities in the multivariate case, not only moments of single variables, but also product moments will be used in the formulas.

We first note useful multivariate forms of Bonferroni's inequalities, given by Meyer [47]. For the bivariate case we consider two classes of events:

$$\{E_{11}, \ldots, E_{1N_1}\} \quad \text{and} \quad \{E_{21}, \ldots, E_{2N_2}\}.$$

Then the probability $P[n_1, n_2]$ that exactly n_1 of the first class and n_2 of the second class of events occur lies between

$$(43) \qquad \sum_{t=n_1+n_2}^{n_1+n_2+2k+1} \sum_{i+j=t} f(i,j;t) \quad \text{and} \quad \sum_{t=n_1+n_2}^{n_1+n_2+2k} f(i,j;t) \quad \text{for any } k > 0,$$

where $f(i,j;t) = (-1)^{t-(n_1+n_2)}\binom{i}{n_1}\binom{j}{n_2}S_{i,j}$ with

$$S_{i,j} = \sum_{t=i+j}^{N_1+N_2} \sum_{g+h=t} \binom{g}{i}\binom{h}{j} P[g,h].$$

The simplest way of deriving multivariate inequalities is to combine univariate inequalities, using the formula

$$\Pr\left[\bigcap_{j=1}^{m} E_j\right] = 1 - \Pr\left[\bigcup_{j=1}^{m} \bar{E}_j\right] \geq 1 - \sum_{j=1}^{m} \Pr[\bar{E}_j]. \tag{44}$$

Taking $E_j \equiv (|X_j - E[X_j]| \leq t_j\sqrt{\text{var}(X_j)})$ and using Chebyshev's inequality (Table 4 of Chapter 33), we obtain

$$\Pr\left[\bigcap_{j=1}^{m}(|X_j - E[X_j]| \leq t_j\sqrt{\text{var}(X_j)})\right] \geq 1 - \sum_{j=1}^{m} t_j^{-2}. \tag{45}$$

Of course, for this formula to be of any use we must have $\sum_{j=1}^{m} t_j^{-2} < 1$, and preferably the sum should be rather small. As is to be expected, this becomes more difficult to ensure as m increases [if $t_1 = t_2 = \cdots = t_m = t$, for instance, the right-hand side of (45) is mt^{-2}]. The inequality (45) (and similar ones that may be obtained from other formulas in Chapter 33) is of rather general applicability. No assumption of independence among the variables is made, nor, indeed, is any specific form of dependence assumed. If independence can be assumed, then

$$\Pr\left[\bigcap_{j=1}^{m}(|X_j - E[X_j]| \leq t_j\sqrt{\text{var}(X_j)})\right] \geq \prod_{j=1}^{m} (1 - t_j^{-2}). \tag{46}$$

Provided that $0 < t_j < 1$ for all j, this inequality gives a nontrivial lower bound. Note that $\prod_{j=1}^{m} (1 - t_j^{-2}) \geq 1 - \sum_{j=1}^{m} t_j^{-2}$, so that (46) gives a larger lower bound than (45), as is to be expected, since the former inequality requires a restriction (namely, independence) in the joint distribution of $X_1, X_2, \ldots, X_m$.

The simplest inequality introducing correlation was obtained by Berge [3] in 1937. This is

$$\Pr\left[\bigcap_{j=1}^{2}\left(\left|\frac{X^j - E[X_j]}{\sigma_j}\right| < t\right)\right] \geq 1 - t^{-2}(1 + \sqrt{1 - \rho^2}), \tag{47}$$

where $\sigma_j = \sqrt{\text{var}(X_j)}$; $\rho = \text{corr}(X_1, X_2)$. This is an improvement over (45) with $m = 2$, $t_1 = t_2 = t$, in that the lower bound $1 - 2t^{-2}$ is replaced by $1 - (1 + \sqrt{1 - \rho^2})t^{-2}$. For $\rho = 1$, when X_1 and X_2 may be regarded as linear functions of each other, we obtain the univariate Chebyshev formula. On the other hand, when $\rho = 0$ we obtain (45), and not the better lower bound $(1 - t^{-2})^2$ corresponding to independence of X_1 and X_2. However, it should be remembered that a zero value for ρ need not imply independence between X_1 and X_2.

Berge [3] obtained the inequality (47) in the following way. We will use the notation $Y_j = (X_j - E[X_j])/\sqrt{\text{var}(X_j)}$ for the standardized X_j variate. Consider the statistic $H = Y_1^2 + Y_2^2 + 2aY_1Y_2$ with $|a| < 1$. If either $|Y_1| \geq t$ or $|Y_2| \geq t$ then $H \geq (1 - a^2)t^2$; and also $H \geq 0$ for all Y_1, Y_2. Hence $E[H] = 2(1 + a\rho) \geq (1 - a^2)t^2 \Pr[\bigcup_{j=1}^{2}(|Y_j| \geq t)]$. Remembering that $\Pr[\bigcap_{j=1}^{2}(|Y_j| < t)] = 1 - \Pr[\bigcup_{j=1}^{2}(|Y_j| \geq t)]$, we have

$$\Pr\left[\bigcap_{j=1}^{2}(|Y_j| < t)\right] \geq 1 - 2(1 + a\rho)(1 - a^2)^{-1}t^{-2}.$$

This formula is valid for any $|a| < 1$. The best value to take for a will minimize the multiplier $2(1 + a\rho)(1 - a^2)^{-1}$. This is effected by taking $a = -\rho^{-1}(1 - \sqrt{1 - \rho^2})$ leading to (47).

By an extension of this argument, Olkin and Pratt [54] in 1958 obtained the inequality

$$\text{(48)} \quad \Pr\left[\bigcap_{j=1}^{m}(|Y_j| < t_j)\right] \geq 1 - m^{-2}\left\{\sqrt{u} + \sqrt{(m-1)\left(m\sum_{j=1}^{m} t_j^{-2} - u\right)}\right\}^2,$$

where $u = \sum_{j=1}^{m} t_j^{-2} + 2\sum\sum_{i<j} \rho_{ij}t_i^{-1}t_j^{-1}$ with $\rho_{ij} = \text{corr}(X_i, X_j)$. For $m = 2$ this gives

$$\text{(49)} \quad \Pr\left[\bigcap_{j=1}^{2}(|Y_j| < t_j)\right] \geq 1 - \tfrac{1}{2}(t_1t_2)^{-2}\{t_1^2 + t_2^2 + (t_1^2 + t_2^2)^2 - 4\rho_{12}^2t_1^2t_2^2\},$$

a result obtained earlier (in 1955) by Lal [35]. Olkin and Pratt [54] point out that it is possible to improve (41), but the necessary calculations will usually be heavy. Godwin [23] generalizes these results to sets of more than two variables.

A further generalization, due to Isii [27a], gives bounds for

$$P = \Pr\left[\bigcap_{j=1}^{2}(-k_1 < X_j < k_2)\right]$$

with $0 < k_1 \leq k_2$.

(a) If $2k_1^2 > 1 - \rho$ and $\frac{1}{2}(k_2 - k_1) \geq \lambda$ with

$$\lambda = \frac{k_1(1 + \rho) + [(1 - \rho^2)(k_1^2 + \rho)]^{1/2}}{2k_1^2 - 1 + \rho}$$

then

$$\text{(50.1)} \qquad P \leq 2\lambda^2(2\lambda^2 + 1 + \rho)^{-1}.$$

(b) If conditions (a) are not satisfied, and also $k_1k_2 \geq 1$ and

$$2(k_1k_2 - 1)^2 \geq 2(1 - \rho^2) + (1 - \rho)(k_2 - k_1)^2$$

then

(50.2)

$$P \leq (k_1 + k_2)^{-2}[(k_2 - k_1)^2 + 4 + \{16(1 - \rho^2) + 8(1 - \rho)(k_2 - k_1)\}^{1/2}].$$

In all other cases there is no universal upper bound for P (other than 1).

An extension of the Gauss-Camp type of inequality (Chapter 33, Table 4), due to Leser [41], is of some interest. He obtained bounds for

$$P = \Pr\left[\sum_{j=1}^{m} \lambda_j^{-2} Y_j^2 \leq m\right].$$

In the univariate case, the Gauss-Camp inequalities are derived on the assumption that the density function of the standardized variable Y is in some sense decreasing as $|Y|$ increases. Leser generalized this by requiring the conditional average of the joint density function $p(y_1, \ldots, y_m)$, given the value of

$$R^2 = \lambda_0^2 \sum_{j=1}^{m} \lambda_j^{-2} y_j^2,$$

(where λ_0^2 is the harmonic mean of $\lambda_1^2, \ldots, \lambda_m^2$) to be a nondecreasing function of R^2 for R^2 less than $m\kappa^2$ (for some $\kappa > 0$.) The inequalities are summarized in Table 3. Note that as m increases, the range $1 \leq \kappa \leq \sqrt{1 - 2m^{-1}}$ becomes narrower. Also, for "really unimodal" distributions κ can be quite large.

TABLE 3

Multivariate Chebyshev-Type Inequalities

κ	λ_0	P
$\kappa \leq 1$	$\lambda_0 \leq 1$	≥ 0
	$\lambda_0 \geq 1$	$\geq 1 - \lambda_0^{-2}$
$1 \leq \kappa \leq \sqrt{1 + 2m^{-1}}$	$\lambda_0 \leq (1 + \frac{1}{2}m)^{-1/m}\kappa$	$\geq (\frac{1}{2}m + 1)(1 - \kappa^{-2})(\lambda_0/\kappa)^m$
	$(1 + \frac{1}{2}m)^{-1/m}\kappa \leq \lambda_0 \leq \kappa$	$\geq 1 - \kappa^{-2}$
	$\lambda_0 \geq \kappa$	$\geq 1 - \lambda_0^{-2}$
$\kappa \geq \sqrt{1 + 2m^{-1}}$	$\lambda_0 \leq (1 + \frac{1}{2}m)^{-1/m}\sqrt{1 + 2m^{-1}}$	$\geq (1 + 2m^{-1})^{-\frac{1}{2}m}\lambda_0^m$
	$(1 + \frac{1}{2}m)^{-1/m}\sqrt{1 + 2m^{-1}} \leq \lambda_0 \leq (1 + \frac{1}{2}m)^{-1/m}\kappa$	$\geq 1 - (1 + \frac{1}{2}m)^{-2/m}\lambda_0^{-2}$
	$(1 + \frac{1}{2}m)^{-1/m}\kappa \leq \lambda_0 \leq \kappa$	$\geq 1 - \kappa^{-2}$
	$\lambda_0 \geq \kappa$	$> 1 - \lambda_0^{-2}$

A very useful discussion of multivariate Chebyshev type inequalities for quite general regions is given by Karlin and Studden [33a].

9. Multivariate Increasing Hazard Rate Distributions

It is not immediately obvious how the concept of increasing hazard rate (IHR), described in Section 7.2 of Chapter 33, can be extended to multivariate distributions. Harris [26] has suggested the following, apparently satisfactory, definition for an IHR joint distribution of m variables $X_1,\ldots,X_m$:

$$\text{(i)}\quad \Pr\left[\bigcap_{j=1}^{m}(X_j > x_j + t)\,\middle|\,\bigcap_{j=1}^{m}(X_j > x_j)\right] \le \Pr\left[\bigcap_{j=1}^{m}(X_j > x_j' + t)\,\middle|\,\bigcap_{j=1}^{m}(X_j > x_j')\right]$$

for all $x_j \ge x_j' \ge 0$ and all $t \ge 0$ and

$$\text{(ii)}\qquad \Pr\left[\bigcap_{j=1}^{m}(X_j > x_j)\,\middle|\,\bigcap_{j=1}^{m}(X_j > x_j')\right]$$

is a nondecreasing function of $x_1',\ldots,x_m'$ for all $x_1,\ldots,x_m$.

It is clear that (ii) is satisfied for any distribution when $x_j' < x_j$ for all j. Condition (ii) must be satisfied for all values of x_j. This condition is called *right corner set increasing*. We note that condition (i) is satisfied with equalities by the multivariate exponential distribution described in Section 5 of Chapter 41. Harris [26] shows:

(a) Any subset of multivariate IHR variables have a multivariate IHR joint distribution (and each marginal distribution is IHR).

(b) The union of two mutually independent sets of variables, each having a joint multivariate IHR distribution, also has a multivariate IHR distribution.

(c) The minima of disjoint subsets of a set of multivariate IHR variables have a joint multivariate IHR distribution.

Harris' definition imposes rather severe restrictions on the distribution function. Goodman [23a] has suggested that a distribution might be termed IHR if the density function $p_{\mathbf{X}}(\mathbf{x})$ is a decreasing function of a *homogeneous* function $g(\mathbf{x})$ of the $\mathbf{x}$'s, and the (univariate) distribution of $g(\mathbf{X})$ is IHR.

10. A Note on Estimation

Weiss and Wolfowitz [79, 80] have developed a method of estimation that they term "generalized maximum likelihood." Among other things, this makes possible a choice between different roots of the formal maximum likelihood equation(s). Essentially, the root to be chosen is that which is nearest to previously specified consistent estimator(s) of the parameter(s). It also applies when the likelihood function is infinite for a number of different values of the parameters.

This method is particularly useful in estimating parameters of mixtures of normal (Chapter 13, Section 7.2) or multinormal (Chapter 35, Section 7) distributions, and also in estimating normal correlation coefficients (Chapter 36, Section 6).

A nonpathological example when these methods can be of practical use is the following.

If X is distributed as a mixture of k multinormal distributions (see Chapter 35, Section 8) with common (though unknown) expected value vector, in known proportions $\omega_1,\omega_2,\ldots,\omega_k\left(\sum_{j=1}^{k}\omega_j = 1\right)$ then the likelihood function of n independent sets of values $\mathbf{X}_j$ ($j = 1,\ldots,n$) is

$$(2\pi)^{-(1/2)mn}\prod_{j=1}^{n}\left(\sum_{i=1}^{k}\omega_i\,|\mathbf{V}_i|^{-1/2}\exp[-\tfrac{1}{2}(\mathbf{X}_j-\boldsymbol{\xi})'\mathbf{V}_i^{-1}(\mathbf{X}_j-\boldsymbol{\xi})]\right)$$

By putting $\hat{\boldsymbol{\xi}} = \mathbf{X}_j$ for some j and taking *one* $|\hat{\mathbf{V}}_i|$ as small as desired (but not varying the other $\hat{\mathbf{V}}_i$'s) the likelihood can be made as large as desired. In such cases the method of maximum likelihood cannot be applied. However, the methods described by Weiss and Wolfowitz [79, 80] can be used.

11. Singular Distributions

It sometimes happens that one or more mathematical relations hold precisely among m random variables $X_1,X_2,\ldots,X_m$. In such cases the joint distribution is said to be *singular*. We shall give no direct discussion of singular distributions, though they will be referred to (e.g., in Section 1 of Chapter 35).

Lukomski [42] has given a systematic treatment of singular distributions. In particular he considered the case when there are just r distinct linear relations among m random variables $X_1,X_2,\ldots,X_m$. He showed that these relations can be derived from the variance covariance matrix of the X's, by replacing an arbitrary row in the left-hand side determinant of each of the r equations

$$\begin{vmatrix} \mu_{r,i} & \mu_{r,r+1} & \cdots & \mu_{r,m} \\ \mu_{r+1,i} & \mu_{r+1,r+1} & \cdots & \mu_{r+1,m} \\ \vdots & \vdots & & \vdots \\ \mu_{m,i} & \mu_{m,r+1} & \cdots & \mu_{m,m} \end{vmatrix} = 0 \quad (i = 1,2,\ldots,r)$$

by $X_i, X_{r+1}, \ldots, X_m$ (the X's being so ordered that

$$\begin{vmatrix} \mu_{r+1,r+1} & \mu_{r+1,r+2} & \cdots & \mu_{r+1,m} \\ \mu_{r+2,r+1} & \mu_{r+2,r+2} & \cdots & \mu_{r+2,m} \\ \vdots & \vdots & & \vdots \\ \mu_{m,r+1} & \mu_{m,r+2} & \cdots & \mu_{m,m} \end{vmatrix} > 0).$$

(In the formulas above, $\mu_{ij} = E[(X_i - E[X_i])(X_j - E[X_j])]$ is the covariance of X_i and X_j.)

REFERENCES

[1] Abazaliev, A. K. J. (1968). Characteristic coefficients of two-dimensional distributions and their applications, *Soviet Mathematics-Doklady*, **9,** 52–56 [English translation; Russian original in *Doklady Akademii Nauk SSSR*, **178** (1968)].

[2] Banys, M. I. (1970). The integral multivariate limit theorem for convergence to a stable limiting distribution, *Lietuvos Matematikos Rinkinys*, **10,** 665–672. (In Russian.)

[3] Berge, P. O. (1938). A note on a form of Tchebycheff's theorem for two variables, *Biometrika*, **29,** 405–406.

[4] Bikelis, A. (1968). On the multivariate characteristic functions, *Lietuvos Matematikos Rinkinys*, **8,** 21–39. (In Russian.)

[5] Bikelis, A. (1968). The asymptotic expansions of the distribution function and of density functions for the sums of independent identically distributed random variables, *Lietuvos Matematikos Rinkinys*, **8,** 405–422.

[6] Bikelis, A. (1970). Two inequalities for the multivariate characteristic function, *Lietuvos Matematikos Rinkinys*, **10,** 1–12. (In Russian.)

[7] Bikelis, A. (1970). Asymptotic expansions of the distribution function of the sum of independent identically distributed non-lattice random vectors, *Lietuvos Matematikos Rinkinys*, **10,** 673–679. (In Russian.)

[8] Bikelis, A. and Mogyoródi, J. (1970). On asymptotic expansion of the convolution of *n* multidimensional distribution functions, *Lietuvos Matematikos Rinkinys*, **10,** 433–443.

[9] Bildikar, Sheela and Patil, G. P. (1968). Multivariate exponential-type distributions, *Annals of Mathematical Statistics*, **39,** 1316–1326.

[10] Chambers, J. M. (1967). On methods of asymptotic approximation for multivariate distributions, *Biometrika*, **54,** 367–383.

[11] Cooper, P. W. (1963). Multivariate extensions of univariate distributions, *IEEE Transactions, Electronic Computers*, **12,** 572–573.

[12] Crofts, A. E. (1969). *An Investigation of Normal Lognormal Distributions*, Technical Report No. 32, Themis Contract, Department of Statistics, Southern Methodist University, Dallas, Texas.

[13] Day, N. E. (1969). Linear and quadratic discrimination in pattern recognition, *IEEE Transactions on Information Theory*, **15,** 419–421.

[14] Dunnage, J. E. A. (1970). The speed of convergence of the distribution functions in the two-dimensional central limit theorem, *Proceedings of the London Mathematical Society, Series 3*, **20,** 33–59.

[15] Dunnage, J. E. A. (1970). On the remainder in the two-dimensional central limit theorem, *Proceedings of the Cambridge Philosophical Society*, **68,** 455–458.

[16] Eagleson, G. K. (1964). Polynomial expansions of bivariate distributions, *Annals of Mathematical Statistics*, **35,** 1208–1215.

[17] Edgeworth, F. Y. (1896). The compound law of error, *Philosophical Magazine, 5th Series*, **41,** 207–215.

[18] Edgeworth, F. Y. (1917). On the mathematical representation of statistical data, *Journal of the Royal Statistical Society, Series A*, **80,** 266–288.

[19] Elderton, W. P. and Johnson, N. L. (1969). *Systems of Frequency Curves*, London: Cambridge University Press.

[20] Farlie, D. J. G. (1960). The performance of some correlation coefficients for a general bivariate distribution, *Biometrika*, **47,** 307–323.

[21] Fréchet, M. (1951). Sur les tableaux de corrélation dont les marges sont données, *Annales de l'Université de Lyon, Section A, Series 3*, **14,** 53–77.

[22] Galton, F. (1877). *Typical Laws of Heredity in Man*, Lecture at the Royal Institution of Great Britain.

[23] Godwin, H. J. (1964). *Inequalities on Distribution Functions*, London: Griffin.

[23a] Goodman, I. (1972). Ph.D. Thesis, Temple University, Philadelphia.

[24] Griffiths, R. C. (1970). Positive definite sequences and canonical correlation coefficients, *Australian Journal of Statistics*, **12,** 162–165.

[25] Guldberg, S. (1920). Application des polynômes d'Hermite à un problème de statistique, *Proceedings of the International Congress of Mathematicians*, Strasbourg, 552–560.

[26] Harris, R. (1970). A multivariate definition for increasing hazard rate distribution functions, *Annals of Mathematical Statistics*, **41,** 713–717.

[27] Hirschfeld, H. O. (Hartley, H. O.) (1935). A connection between correlation and contingency, *Proceedings of the Cambridge Philosophical Society*, **31,** 520–524.

[27a] Isii, K. (1959) On a method for generalizations of Tchebycheff's inequality, *Annals of the Institute of Statistical Mathematics, Tokyo*, **10,** 65–88.

[28] Jensen, D. R. (1971). A note on positive dependence and the structure of bivariate distributions, *SIAM Journal of Applied Mathematics*, **20,** 749–752.

[29] Johnson, N. L. (1949). Bivariate distributions based on simple translation systems, *Biometrika*, **36,** 297–304.

[30] Jones, R. M. and Miller, K. S. (1966). On the multivariate lognormal distribution, *Journal of Industrial Mathematics*, **16,** 63–76.

[31] Jørgensen, N. R. (1916). *Undersøgelser over Frequensflader of Correlation.* Copenhagen: Busch.

[32] Kalinauskaité, N. (1970). On some expansions for the multidimensional stable densities with parameter $\alpha > 1$, *Lietuvos Matematikos Rinkinys*, **10,** 490–495. (In Russian.)

[33] Kalinauskaité, N. (1970). On some expansions for the multidimensional symmetric stable densities, *Lietuvos Matematikos Rinkinys*, **10,** 726–731. (In Russian.)

[33a] Karlin, S. and Studden, W. J. (1966) *Tchebycheff Systems: With Applications in Analysis and Statistics*, New York: John Wiley & Sons.

[34] Koopmans, L. H. (1969). Some simple singular and mixed probability distributions, *American Mathematical Monthly*, **76,** 297–299.

[35] Lal, D. N. (1955). A note on a form of Tchebycheff's inequality for two or more variables, *Sankhyā*, **15,** 317–320.

[36] Lancaster, H. O. (1957). Some properties of the bivariate normal distribution considered in the form of a contingency table, *Biometrika*, **44,** 289–292.

[37] Lancaster, H. O. (1958). The structure of bivariate distributions, *Annals of Mathematical Statistics*, **29,** 719–736.

[38] Lancaster, H. O. (1963). Correlations and canonical forms of bivariate distributions, *Annals of Mathematical Statistics*, **34,** 532–538.

[39] Lancaster, H. O. (1965). Symmetry in multivariate distributions, *Australian Journal of Statistics*, **7,** 115–126.

[40] Lancaster, H. O. (1969). *The Chi-squared Distribution*, New York: John Wiley and Sons, Inc.

[41] Leser, C. E. V. (1942). Inequalities for multivariate frequency distributions, *Biometrika*, **32,** 284–293.

[42] Lukomski, J. (1939). On some properties of multidimensional distributions, *Annals of Mathematical Statistics*, **10,** 236–246.

[43] Mamatkulov, K. K. (1970). Quadratic estimation of the density function of bivariate lognormal density function from a sample, *Izvestia Akademii Nauk Uzbek SSR, Seria Fiziko-Matematicheskikh Nauk*, **14,** No. 1, 17–20. (In Russian.)

[44] Mardia, K. V. (1970). A translation family of bivariate distributions and Fréchet's bounds, *Sankhyā, Series A*, **32,** 119–121.

[45] Mardia, K. V. (1970). *Families of Bivariate Distributions*, London: Griffin.

[46] Maung, K. (1941). Measurement of association in a contingency table with special reference to the pigmentation of hair and eye colours of Scottish school children, *Annals of Eugenics, London*, **11,** 189–223.

[46a] Meixner, J. (1934). Orthogonale Polynomsysteme mit einer besonderen Gestalt des erzeugenden Funktion, *Journal of the London Mathematical Society*, **9,** 6–13.

[47] Meyer, R. M. (1969). Note on a "multivariate" form of Bonferroni's inequalities, *Annals of Mathematical Statistics*, **40,** 692–693.

[48] Mihaïla, I. M. (1968). Development of the trivariate frequency function in Gram-Charlier series, *Revue Roumaine de Mathématiques Pures et Appliquées*, **13,** 803–813.

[49] Morgenstern, D. (1956). Einfache Beispiele zweidimensionaler Verteilungen, *Mitteilingsblatt für Mathematische Statistik*, **8,** 234–235.

[50] Mostafa, M. D. and Mahmoud, M. W. (1964). On the problem of estimation for the bivariate lognormal distribution, *Biometrika*, **51,** 522–527.

[51] Narumi, S. (1923). On the general forms of bivariate frequency distributions which are mathematically possible when regression and variation are subjected to limiting conditions, *Biometrika*, **15**, 77–88, 209–221.

[52] Nataf, A. (1962). Détermination des distributions dont les marges sont données, *Comptes Rendus de l'Académie des Sciences, Paris*, **225**, 42–43.

[53] Neyman, J. (1926). Further notes on non-linear regression, *Biometrika*, **18**, 257–262.

[54] Olkin, I. and Pratt, J. W. (1958). A multivariate Tchebycheff inequality, *Annals of Mathematical Statistics*, **29**, 226–234.

[55] Paulauskas, V. (1970). On the multidimensional central limit theorem, *Lietuvos Matematikos Rinkinys*, **10**, 783–789. (In Russian.)

[56] Pearson, K. (1905). On the general theory of skew correlation and nonlinear regression, *Drapers' Company Research Memoirs, Biometric Series*, **2**.

[57] Pearson, K. (1923). On non-skew frequency surfaces, *Biometrika*, **15**, 231–244.

[58] Pearson, K. (1923). Notes on skew frequency surfaces, *Biometrika*, **15**, 222–230.

[59] Plackett, R. L. (1965). A class of bivariate distributions, *Journal of the American Statistical Association*, **60**, 516–522.

[60] Pretorius, S. J. (1930). Skew bivariate frequency surfaces, examined in the light of numerical illustrations, *Biometrika*, **22**, 109–223.

[61] Rhodes, E. C. (1923). On a certain skew correlation surface, *Biometrika*, **14**, 355–377.

[62] Risser, R. (1945). Sur l'équation caractéristique des surfaces de probabilité, *Comptes Rendus, Académie des Sciences, Paris*, **220**, 31–32.

[63] Risser, R. (1947). D'un certain mode de recherche des surfaces de probabilité, *Comptes Rendus, Académie des Sciences, Paris*, **225**, 1266–1268.

[64] Risser, R. (1950). Calcul des constantes de certaines surfaces de distribution, *Bulletin des Actuaires Français*, No. 191, 141–232.

[65] Risser, R. and Traynard, C. E. (1957). *Les Principes de la Statistique Mathématique*, **2** (Part 2), Paris: Gauthier Villars.

[66] Roux, J. J. J. (1971). A characterization of a multivariate distribution, *South African Statistical Journal*, **5**, 27–36.

[67] Sagrista, S. N. (1952). On a generalization of Pearson's curves to the two-dimensional case, *Trabajos de Estadìstica*, **3**, 273–314. (In Spanish.)

[68] Sarmanov, O. V. (1958). Maximal correlation coefficient. Symmetrical case, *Doklady Akademii Nauk SSSR*, **120**, 715–718. (In Russian.)

[69] Sarmanov, O. V. (1965). On the method of characteristic coefficients, *Soviet Mathematics-Doklady*, **6**, 1083–1091. [English translation, Russian original in *Doklady Akademii Nauk SSSR*, **163** (1956).]

[70] Sarmanov, O. V. (1966). Generalized normal correlation and two-dimensional Fréchet classes, *Soviet Mathematics—Doklady*, **7,** 596–599 [English translation; Russian original in *Doklady Akademii Nauk SSSR*, **108** (1966)].

[71] Sarmanov, O. V. and Bratoeva, Z. N. (1967). Probabilistic properties of bilinear expansions in Hermite polynomials, *Teoriya Veroyatnostei i ee Primeneniya*, **12,** 520–531. (In Russian; English translation pp. 470–481.)

[72] Sazonov, V. V. (1967). On the rate of convergence in the multidimensional central limit theorem, *Teoriya Veroyatnostei i ee Primeneniya*, **12,** 82–95. (In Russian; English translation pp. 77–89.)

[73] Seshadri, V. and Patil, G. P. (1964). A characterization of a bivariate distribution by the marginal and the conditional distributions of the same component, *Annals of the Institute of Statistical Mathematics, Tokyo*, **15,** 215–221.

[74] Steyn, H. S. (1960). On regression properties of multivariate probability functions of Pearson's types, *Proceedings of the Royal Academy of Sciences, Amsterdam*, **63,** 302–311.

[75] Uven, M. J. van (1925–1926). On treating skew correlation, *Proceedings of the Royal Academy of Sciences, Amsterdam*, **28,** 797–811, 919–935; **29,** 580–590.

[76] Uven, M. J. van (1929). Skew correlation between three and more variables, I–III, *Proceedings of the Royal Academy of Sciences, Amsterdam*, **32,** 793–807, 995–1007, 1085–1103.

[77] Uven, M. J. van (1947–1948). Extension of Pearson's probability distributions to two variables, I–IV. *Proceedings of the Royal Academy of Sciences, Amsterdam*, **50,** 1063–1070, 1252–1264; **51,** 41–52, 191–196.

[78] Wani, J. K. (1968). On the linear exponential family, *Proceedings of the Cambridge Philosophical Society*, **64,** 481–485.

[79] Weiss, L. and Wolfowitz, J. (1966). Generalized maximum likelihood estimators, *Theory of Probability and Its Applications*, **11,** 58–81. (Reprinted from Teoriya Veroyatnostei i ee Primeneniya, **11,** 68–93.)

[80] Weiss, L. and Wolfowitz, J. (1967). Maximum probability estimators, *Annals of the Institute of Statistical Mathematics, Tokyo*, **19,** 193–206.

[81] Wicksell, S. D. (1917). On the genetic theory of frequency, *Arkiv for Matematik Astronomi och Fysik*, **12,** No. 20.

[82] Wicksell, S. D. (1923). Contributions to the analytical theory of sampling, *Arkiv for Matematik Astronomi och Fysik*, **17,** No. 19.

[83] Yule, G. U. (1897). On the significance of Bravais' formulae for regression in the case of skew correlation, *Proceedings of the Royal Society of London*, **60,** 477–489.

[84] Zolotarev, V. M. (1966). A multidimensional analogue of the Berry-Esseen inequality for sets with bounded diameter, *Theory of Probability and Its Applications*, **11,** 447–454.

35

Multinormal Distributions

1. Introduction and Genesis

A word of explanation is needed about the reason for discussing the general multinormal distribution here and the simpler cases of bivariate and trivariate normal distributions in Chapter 36. The reason is that there is a considerably greater volume of results on latter special cases than on general multinormal distributions. Their review therefore must be more detailed. By examining the general distribution first we are able to concentrate on details specific to the bivariate and trivariate cases in Chapter 36. Historical remarks will be found in Section 2 of Chapters 34 and 36.

If $U_1, U_2, \ldots, U_m$ are independent unit normal variables their joint density is

$$p_{\mathbf{U}}(\mathbf{u}) = (2\pi)^{-(1/2)m} \exp\left[-\tfrac{1}{2}\sum_{j=1}^{m} u_j^2\right] = (2\pi)^{-(1/2)m} \exp(-\tfrac{1}{2}\mathbf{u}'\mathbf{u}).$$

Applying the nonsingular linear transformation to $\mathbf{Y}' = (Y_1, \ldots, Y_m)$,

$$\mathbf{U}' = \mathbf{Y}'\mathbf{H}', \qquad \text{with} \quad |\mathbf{H}| \neq 0,$$

we find that $\mathbf{Y}$ has joint density function

$$\begin{aligned} p_{\mathbf{Y}}(\mathbf{y}) &= (2\pi)^{-(1/2)m}\, |\mathbf{H}| \exp[-\tfrac{1}{2}\mathbf{y}'\mathbf{H}'\mathbf{H}\mathbf{y}] \\ &= (2\pi)^{-(1/2)m}\, |\mathbf{A}|^{1/2} \exp[-\tfrac{1}{2}\mathbf{y}'\mathbf{A}\mathbf{y}] \qquad \text{with} \quad \mathbf{A} = \mathbf{H}'\mathbf{H}, \end{aligned}$$

so that **A** is positive definite. This is a special case of a multinormal distribution. The variance-covariance matrix of **Y** is (since $E[\mathbf{Y}] = 0$),

$$\begin{aligned}\mathbf{Var}(\mathbf{Y}) = E[\mathbf{YY}'] &= E[\mathbf{H}^{-1}\mathbf{UU}'\mathbf{H}'^{-1}]\\ &= \mathbf{H}^{-1}E[\mathbf{UU}']\mathbf{H}'^{-1}\\ &= \mathbf{H}^{-1}\mathbf{H}'^{-1}\\ &= \mathbf{A}^{-1}.\end{aligned}$$

If we consider the joint distribution of $\mathbf{Z}'$, where $\mathbf{U}' + \boldsymbol{\zeta}' = \mathbf{Z}'\mathbf{H}'$, we obtain

$$p_{\mathbf{Z}}(\mathbf{z}) = (2\pi)^{-(1/2)m}\,|\mathbf{A}|^{\frac{1}{2}}\exp[-\tfrac{1}{2}(\mathbf{z} - \boldsymbol{\zeta})'\mathbf{A}(\mathbf{z} - \boldsymbol{\zeta})]$$

with a more general form [in fact, the most general form; see text following equation (3′)] of multinormal distribution. Here $E[\mathbf{Z}] = \boldsymbol{\zeta}$, $\mathbf{Var}(\mathbf{Z}) = \mathbf{A}^{-1}$ $(= \mathbf{Var}(\mathbf{Y}))$.

The multivariate normal distribution is a limiting form of the multinomial distribution (see Chapter 11). If $X_1, X_2, \ldots, X_{m+1}$ have a joint multinomial distribution with parameters $n, p_1, p_2, \ldots, p_{m+1}\left(\sum_{j=1}^{m+1} p_j = 1\right)$, then the limiting joint distribution, as $n \to \infty$ of the standardized variables

$$Y_j = (X_j - np_j)(np_j(1 - p_j))^{-1/2} \qquad (j = 1, \ldots, m + 1)$$

is multinormal. Note that only the m variables $Y_1, Y_2, \ldots, Y_m$ are included here. The joint distribution of $Y_1, \ldots, Y_m, Y_{m+1}$ is a *singular* multinormal distribution (see Section 2) because there is a fixed linear relation among the $(m + 1)$ variables $(\Sigma Y_j\{p_j(1 - p_j)\}^{1/2} = 0)$.

The multivariate normal distribution is also the limiting joint distribution (as $n \to \infty$) of standardized variables corresponding to $S_1, S_2, \ldots, S_m$ where

$$S_j = \sum_{i=1}^{n} X_{ji}$$

and $(X_{1i}, \ldots, X_{mi})$ have the same joint distribution with finite means and variances for all $i = 1, 2, \ldots, m$; and $(X_{1i}, \ldots, X_{mi})$ and $(X_{1i'}, \ldots, X_{mi'})$ are mutually independent if $i \neq i'$. (See also Chapter 34, Section 4.)

2. Definition and Moments

The random variables $X_1, X_2, \ldots, X_m$ have a *joint multinormal distribution* if their joint probability density function can be written in the form

$$(1)\quad p_{X_1,\ldots,X_m}(x_1, \ldots, x_m) = C\exp[-(\text{positive definite quadratic form in } x_1, \ldots, x_m)]$$

with C an appropriate constant. Writing the exponent as $-\frac{1}{2}(\mathbf{x}-\boldsymbol{\xi})'\mathbf{A}(\mathbf{x}-\boldsymbol{\xi})$ where $\mathbf{A}$ is a real symmetric positive definite matrix, we see that C must be a function of $\boldsymbol{\xi}$ and $\mathbf{A}$. In order to find the value of C we evaluate the joint moment generating function of the X's:

$$\Phi(t_1, \ldots, t_m) = E[e^{\mathbf{t}'\mathbf{X}}].$$

We have

$$\Phi(\mathbf{t}) = C\int_{-\infty}^{\infty}\cdots\int_{-\infty}^{\infty}\exp[-\tfrac{1}{2}(\mathbf{x}-\boldsymbol{\xi})'\mathbf{A}(\mathbf{x}-\boldsymbol{\xi}) + \mathbf{t}'\mathbf{x}]\,d\mathbf{x}.$$

Making the transformation $\mathbf{y} = \mathbf{x} - \boldsymbol{\xi}$ we find

$$\Phi(\mathbf{t}) = Ce^{\mathbf{t}'\boldsymbol{\xi}}\int_{-\infty}^{\infty}\cdots\int_{-\infty}^{\infty}\exp[-\tfrac{1}{2}\mathbf{y}'\mathbf{A}\mathbf{y} + \mathbf{t}'\mathbf{y}]\,d\mathbf{y}.$$

Since $\mathbf{A}$ is positive definite, $\mathbf{A} = \mathbf{H}'\mathbf{H}$ with, of course, $|\mathbf{A}| = |\mathbf{H}|^2$. Making the transformation $\mathbf{z}' = \mathbf{y}'\mathbf{H}'$ (with Jacobian $\partial(\mathbf{z}')/\partial(\mathbf{y}') = |\mathbf{H}|$), we have

$$(2)\quad \Phi(\mathbf{t}) = C\,|\mathbf{H}|^{-1}\,e^{\mathbf{t}'\boldsymbol{\xi}}\int_{-\infty}^{\infty}\cdots\int_{-\infty}^{\infty}\exp[-\tfrac{1}{2}\mathbf{z}'\mathbf{z} + \mathbf{t}'\mathbf{H}'^{-1}\mathbf{z}]\,d\mathbf{z}$$

$$= C\,|\mathbf{H}|^{-1}\,e^{\mathbf{t}'\boldsymbol{\xi}}\int_{-\infty}^{\infty}\cdots\int_{-\infty}^{\infty}\exp\left[-\tfrac{1}{2}\sum_{j=1}^{m}(z_j^2 + 2b_jz_j)\right]dz_1\cdots dz_m$$

with $\mathbf{b}' = (b_1, \cdots, b_m) = \mathbf{t}'\mathbf{H}'^{-1}$. Since $z_j^2 + 2b_jz_j = (z_j + b_j)^2 - b_j^2$ and $\mathbf{b}'\mathbf{b} = \sum_{j=1}^{m} b_j^2$, (2) can be written as

$$\Phi(\mathbf{t}) = C\,|\mathbf{H}|^{-1}\exp\{\mathbf{t}'\boldsymbol{\xi} + \tfrac{1}{2}\mathbf{b}'\mathbf{b}\}\prod_{j=1}^{m}\int_{\infty}^{\infty}\exp[-\tfrac{1}{2}(z_j + b_j)^2]\,dz_j$$

$$= C\,|\mathbf{H}|^{-1}(2\pi)^{(1/2)m}\exp\{\mathbf{t}'\boldsymbol{\xi} + \tfrac{1}{2}\mathbf{b}'\mathbf{b}\}.$$

Finally, since $|\mathbf{H}| = |\mathbf{A}|^{1/2}$ and $\mathbf{b}'\mathbf{b} = \mathbf{t}'\mathbf{H}'^{-1}\mathbf{H}^{-1}\mathbf{t} = \mathbf{t}'\mathbf{A}^{-1}\mathbf{t}$, we have

$$\Phi(\mathbf{t}) = C\,|\mathbf{A}|^{-1/2}(2\pi)^{(1/2)m}\exp(\mathbf{t}'\boldsymbol{\xi} + \tfrac{1}{2}\mathbf{t}'\mathbf{A}^{-1}\mathbf{t}).$$

Since $\Phi(\mathbf{0}) = 1$, it follows that $C = |\mathbf{A}|^{1/2}(2\pi)^{-(1/2)m}$, so that the joint density function is

$$(3)\qquad p_{\mathbf{X}}(\mathbf{x}) = (2\pi)^{-(1/2)m}\,|\mathbf{A}|^{1/2}\exp\{-\tfrac{1}{2}(\mathbf{x}-\boldsymbol{\xi})'\mathbf{A}(\mathbf{x}-\boldsymbol{\xi})\}.$$

$$(4)\qquad \Phi(\mathbf{t}) = \exp(\mathbf{t}'\boldsymbol{\xi} + \tfrac{1}{2}\mathbf{t}'\mathbf{A}^{-1}\mathbf{t}).$$

From (4)

$$(5.1)\qquad E[\mathbf{X}] = \boldsymbol{\xi};$$

variance-covariance matrix of $\mathbf{X}$ is $\mathbf{A}^{-1}$, or symbolically

$$(5.2)\qquad \mathbf{V}(\mathbf{X}) = \mathbf{A}^{-1}.$$

(Sometimes the notation **Var(X)** is used; sometimes **(X)** is omitted.) In terms of **V**, (3) becomes

$$(3)' \qquad p_{\mathbf{X}}(\mathbf{x}) = (2\pi)^{-(1/2)m}\,|\mathbf{V}|^{-1/2}\exp\{-\tfrac{1}{2}(\mathbf{x}-\boldsymbol{\xi})'\mathbf{V}^{-1}(\mathbf{x}-\boldsymbol{\xi})\}.$$

Furthermore, all cumulants and cross-cumulants of order higher than 2 are zero.

Note that since we can *always* find **H** such that $\mathbf{A} = \mathbf{H}'\mathbf{H}$, *any* multinormal distribution can be constructed as the joint distribution of linear functions of independent normal variables, as described in Section 1.

A derivation of the value of C, by Todhunter [124] in 1869, is of some historical interest.

If **A** is only positive *semi*-definite (i.e., $|\mathbf{A}| = 0$), the joint distribution of $X_1, X_2, \ldots, X_m$ is called *singular multinormal.*

Note that since $(\mathbf{X}-\boldsymbol{\xi})'\mathbf{A}(\mathbf{X}-\boldsymbol{\xi}) = \mathbf{Z}'\mathbf{Z}$ with $\mathbf{Z}' = (\mathbf{Z}_1, \ldots, \mathbf{Z}_m)$ comprised of independent unit normal variables, this quadratic form is distributed as χ^2 with m degrees of freedom.

The entropy of the distribution (3) (with $\mathbf{V} = \mathbf{A}^{-1}$) is

$$-E[\log p_{\mathbf{X}}(\mathbf{X})] = \tfrac{1}{2}m \log 2\pi + \tfrac{1}{2}\log|\mathbf{V}| + \tfrac{1}{2}m.$$

Rao [98] has shown that this is the maximum entropy possible for any random vector of m dimensions with specified variance covariance matrix **V**. No other distribution attains this maximum.

3. Other Properties

From the form of the density function (1) it is clear that if any subset—$X_1, \ldots, X_s$, say—of variables is eliminated by "integrating out," the remaining variables $X_{s+1}, X_{s+2}, \ldots, X_m$ have a joint density function of the same form. This means that $X_{s+1}, X_{s+2}, \ldots, X_m$ also have a joint multinormal distribution. In particular, each variable has a normal distribution. The parameters of each distribution are given by (5.1) and (5.2).

The other parameters (correlations) of the joint distribution of $X_{s+1}, X_{s+2}, \ldots, X_m$ could also be found from (5.1) and (5.2). The following argument, however, obtains concise formulas for the parameters, and also derives the conditional joint distribution of $X_1, \ldots, X_s$, given $X_{s+1}, X_{s+2}, \ldots, X_m$.

We partition the matrix **A** at the sth row and column to give

$$(6) \qquad \mathbf{A} = \begin{pmatrix} \mathbf{A}_{11} & \mathbf{A}_{12} \\ \mathbf{A}_{21} & \mathbf{A}_{22} \end{pmatrix}.$$

(Note that $\mathbf{A}_{21} = \mathbf{A}'_{12}$.) The similarly partitioned $m \times m$ matrix

$$\mathbf{C} = \begin{pmatrix} \mathbf{I} & \mathbf{0} \\ -\mathbf{A}_{21}\mathbf{A}_{11}^{-1} & \mathbf{I} \end{pmatrix}$$

satisfies the equation

$$\mathbf{CAC}' = \begin{pmatrix} \mathbf{A}_{11} & \mathbf{0} \\ \mathbf{0} & \mathbf{A}_{22} - \mathbf{A}_{21}\mathbf{A}_{11}^{-1}\mathbf{A}_{12} \end{pmatrix}. \tag{7}$$

Hence making the transformation

$$(\mathbf{x} - \boldsymbol{\xi})' = \mathbf{y}'\mathbf{C},$$

we have

$$(\mathbf{x} - \boldsymbol{\xi})'\mathbf{A}(\mathbf{x} - \boldsymbol{\xi}) = \mathbf{y}'\mathbf{CAC}'\mathbf{y} = \mathbf{y}'_{(1)}\mathbf{A}_{11}\mathbf{y}_{(1)} + \mathbf{y}'_{(2)}\mathbf{D}\mathbf{y}_{(2)},$$

where $\mathbf{y}'_{(1)} = (y_1,y_2,\ldots,y_s)$; $\mathbf{y}'_{(2)} = (y_{s+1},\ldots,y_m)$ and

$$\mathbf{D} = \mathbf{A}_{22} - \mathbf{A}_{21}\mathbf{A}_{11}^{-1}\mathbf{A}_{12}.$$

The joint density function of the variates $\mathbf{Y}' = (Y_1, Y_2,\ldots, Y_m)$ defined by $(\mathbf{X} - \boldsymbol{\xi})' = \mathbf{Y}'\mathbf{C}$ is

$$p_{\mathbf{Y}'}(\mathbf{y}_{(1)},\mathbf{y}_{(2)}) = \frac{|\mathbf{A}_{11}|^{1/2}}{(2\pi)^{(1/2)s}} \exp(-\tfrac{1}{2}\mathbf{y}'_{(1)}\mathbf{A}_{11}\mathbf{y}'_{(1)}) \frac{|\mathbf{D}|^{1/2}}{(2\pi)^{\frac{1}{2}(m-s)}} \exp(-\tfrac{1}{2}\mathbf{y}'_{(2)}\mathbf{D}\mathbf{y}_{(2)}), \tag{8}$$

since $|\mathbf{A}| = |\mathbf{A}_{11}|\,|\mathbf{D}|$ [from (7), noting that $|\mathbf{C}| = 1$].

It follows from (8) that the sets $\mathbf{Y}'_{(1)} = (Y_1,\ldots,Y_s)$ and $\mathbf{Y}'_{(2)} = (Y_{s+1},\ldots, Y_m)$ are independent of each other, and each has a joint multinormal distribution. Examining $\mathbf{C}$ more closely we see that $\mathbf{Y}'_{(2)} = \mathbf{X}'_{(2)}$, i.e., $Y_j = X_j - \xi_j$ for $j = s+1, s+2,\ldots,m,$ while $\mathbf{Y}'_{(1)} = (\mathbf{X}_{(1)} - \boldsymbol{\xi}_{(1)})' + (\mathbf{X}_{(2)} - \boldsymbol{\xi}_{(2)})'\mathbf{A}_{21}\mathbf{A}_{11}^{-1}$. Thus we can restate our results as follows:

(i) $X_{s+1},X_{s+2},\ldots,X_m$ have a joint multinormal distribution with expected values $\xi_{s+1},\ldots,\xi_m$ and variance-covariance matrix $(\mathbf{A}_{22} - \mathbf{A}_{21}\mathbf{A}_{11}^{-1}\mathbf{A}_{12})^{-1}$.

(ii) The conditional joint distribution of $\mathbf{X}'_{(1)} = (X_1,X_2,\ldots,X_s)$, given $X_{s+1},\ldots,X_m$ is multinormal with expected values

$$\boldsymbol{\xi}'_{(1)} - (\mathbf{X}_{(2)} - \boldsymbol{\xi}_{(2)})'\mathbf{A}_{21}\mathbf{A}_{11}^{-1} \tag{9}$$

and variance-covariance matrix $\mathbf{A}_{11}^{-1}$. Formula (9) shows that the regression of each of $X_1,X_2,\ldots,X_s$ on the set $\mathbf{X}'_{(2)} = (X_{s+1},\ldots,X_m)$ is *linear* and *homoscedastic* (since $\mathbf{A}_{11}^{-1}$ does not depend on $\mathbf{X}'_{(2)}$).

From (8) it can be seen that the joint distribution of any linear functions of the X's will be a (singular or nonsingular) multinormal distribution.

Šidák [109] (see also Dunn [31]) has shown that if $X_1,X_2,\ldots,X_m$ have a joint multinormal distribution, then

$$\Pr\left[\bigcap_{j=1}^{m}(|X_j - \xi_j| \leq c_j)\right] \geq \prod_{j=1}^{m} \Pr[|X_j - \xi_j| \leq c_j] \tag{10}$$

for any set of positive constants $c_1,c_2,\ldots,c_m$.

Scott [106] (see also Khatri [68], for a particular case) has proved the inequality obtained by replacing $\leq c_j$ in (10) by $\geq c_j$ (twice).

Gupta [46] has proved the more general results that if $\mathcal{C}_1$ and $\mathcal{C}_2$ are convex sets, symmetrical about $(\xi_1,\ldots,\xi_{m_1}),(\xi_{m_1+1},\ldots,\xi_m)$ in the space of $\mathbf{X}_{(1)} = (X_1,\ldots,X_{m_1}),\mathbf{X}_{(2)} = (X_{m_1+1},\ldots,X_m)$ respectively, then

$$\Pr[(\mathbf{X}_{(1)} \in \mathcal{C}_1) \cap (\mathbf{X}_{(2)} \in \mathcal{C}_2)] \geq \Pr[\mathbf{X}_{(1)} \in \mathcal{C}_1]\Pr[\mathbf{X}_{(2)} \in \mathcal{C}_2] \tag{11.1}$$

$$\Pr[(\mathbf{X}_{(1)} \in \bar{\mathcal{C}}_1) \cap (\mathbf{X}_{(2)} \in \bar{\mathcal{C}}_2)] \geq \Pr[\mathbf{X}_{(1)} \in \bar{\mathcal{C}}_1]\Pr[\mathbf{X}_{(2)} \in \bar{\mathcal{C}}_2] \tag{11.2}$$

where $\bar{\mathcal{C}}_j$ is the complement of $\mathcal{C}_j$ $(j = 1,2)$.

Slepian [112] showed that (for any $c_1,c_2,\ldots,c_m$) the derivative of $\Pr\left[\bigcap_{j=1}^{m}(X_j - \xi_j \leq c_j)\right]$ with respect to $\rho_{ii'}$ is nonnegative for all i,i'. He used (14) (below) to establish this result. Jogdeo [56] showed that if ρ_{1i} $(= \rho_{i1})$ are increased by a multiplier λ (other ρ's remaining the same) then

$$\frac{d}{d\lambda}\Pr\left[\bigcap_{j=1}^{m}(|X_j - \xi_j| \leq c_j)\right] \geq 0.$$

For the particular (symmetric) case, when all correlation coefficients are equal and positive, Tong [125] has obtained inequalities between certain probabilities relating to different numbers of variables. In particular, for $m \geq k \geq 2$

$$\begin{aligned}\Pr\left[\bigcap_{j=1}^{m}(X_j - \xi_j \leq d\sigma_j)\right] &\geq \left\{\Pr\left[\bigcap_{j=1}^{k}(X_j - \xi_j \leq d\sigma_j)\right]\right\}^{m/k} \\ &\geq \Phi(d) + \left\{\Pr\left[\bigcap_{j=1}^{2}(X_j - \xi_j \leq d\sigma_j)\right] - [\Phi(d)]^2\right\}^{m/2}.\end{aligned} \tag{11.3}$$

The same inequalities hold (for $d > 0$) with $X_j - \xi_j$ replaced by $|X_j - \xi_j|$.

4. Evaluation of Multinormal Probabilities

In this section we consider, for the most part, *standardized* multinormal distributions, that is, density functions of the form

$$p_{\mathbf{X}}(\mathbf{x}) = \frac{|\mathbf{R}|^{1/2}}{(2\pi)^{(1/2)m}} \exp[-\tfrac{1}{2}\mathbf{x}'\mathbf{R}^{-1}\mathbf{x}], \tag{12}$$

where $\mathbf{X}' \equiv (X_1,X_2,\ldots,X_m);\mathbf{x}' = (x_1,x_2,\ldots,x_m)$ and $\mathbf{R}$ is the *correlation* matrix of $\mathbf{X}$. The notation $Z_m(\mathbf{x};\mathbf{R})$ is sometimes used for this function. We shall do so.

Generalizing the univariate probability integral $\Phi(h)$, we define

$$\begin{aligned}(13)\quad \Phi_m(h_1,\ldots,h_m;\mathbf{R}) &= \Pr\left[\bigcap_{j=1}^{m}(X_j \le h_j)\right]\\ &= \frac{|\mathbf{R}|^{1/2}}{(2\pi)^{(1/2)m}} \int_{-\infty}^{h_m}\cdots\int_{-\infty}^{h_1} \exp[-\tfrac{1}{2}\mathbf{x}'\mathbf{R}^{-1}\mathbf{x}]\, dx_1 \cdots dx_m.\end{aligned}$$

(Later, as in Section 2, we shall use $d\mathbf{x}$ as an abbreviation for $dx_1\ldots dx_m$). This definition applies also when $\mathbf{R}$ is a variance-covariance matrix. There does not appear to be any simple procedure for evaluating (13) in the general cases. Reduction formulas developed by Plackett [96], Steck [115], and John [57], which are described below, are unfortunately somewhat laborious in practice (even with assistance of electronic computers) when ρ is greater than $\frac{1}{2}$. There are, however, simplifications in certain special cases, which will be described later in this section. Also, some calculations are practicable for smaller values of m. Those for $m = 4$ are described in Section 5 of this chapter; detailed discussion of the cases $m = 2$ and $m = 3$ appears in Chapter 36.

We shall not discuss calculation of multinormal probabilities other than those of form (13). Evaluation of multinormal probabilities over convex polyhedra has been described by John [59]. (See also van de Vaart [127, 128].) The special case of convex polygons will be discussed in Chapter 36 (Section 4).

Evaluation of multinormal integrals by some form of Monte Carlo technique has been studied recently by Escoufier [36] and Abbe [1], among others. In [36], some simplification in evaluation of integrals over regions bounded by planes ($\mathbf{a}'\mathbf{X} = \mathbf{c}$) is obtained by transforming to independent variables $\mathbf{Z}$ as in Section 2. In [1] the use of varying sampling rates in different parts of the region of integration is discussed.

4.1. *Reduction Formulas*

Plackett [96] based his reduction formula on the differential equation:

$$\frac{\partial Z_m(\mathbf{x};\mathbf{R})}{\partial \rho_{ij}} = \frac{\partial^2 Z_m(\mathbf{x};\mathbf{R})}{\partial x_i\, \partial x_j}. \tag{14}$$

Suppose that $\Phi(\mathbf{h};\mathbf{R})$ can be evaluated for $\mathbf{R} = \mathbf{R}_0$. Then it follows from (14) that

$$\Phi_m(\mathbf{h};\mathbf{R}) = \Phi(\mathbf{h},\mathbf{R}_0) + \sum_{i<j}\sum \int_{\rho_{ij0}}^{\rho_{ij}} \frac{\partial \Phi_m(\mathbf{h};t\mathbf{R} + (1-t)\mathbf{R}_0)}{\partial \lambda_{ij}}\, d\lambda_{ij} \tag{15}$$

with

$$\lambda_{ij} = t\rho_{ij} + (1-t)\rho_{ij0}$$

(where ρ_{ij0} denotes the i,jth element of $\mathbf{R}_0$).

Plackett also derived the formula

$$\frac{\partial \Phi_m(\mathbf{h};\mathbf{R})}{\partial \rho_{12}} = Z_2(h_1,h_2;\rho_{12})\Phi_{m-2}(h_3 - \bar{h}_3,\ldots,h_m - \bar{h}_m;\mathbf{V}_{(2)}), \tag{16}$$

where

$$\bar{h}_j = \frac{(\rho_{1j} - \rho_{2j}\rho_{12})h_1 + (\rho_{2j} - \rho_{1j}\rho_{12})h_2}{\sqrt{(1-\rho_{12}^2)}} \qquad (j \neq 1,2)$$

and $\mathbf{V}_{(2)}$ is the conditional variance-covariance matrix of $X_3,\ldots,X_m$ given X_1 and X_2. (See also Poznyakov [96a].)

Steck [115] noted that if h_ih_j is not negative for any pair (i,j), then

$$\Phi(\mathbf{h};\mathbf{R}) = \sum_{j=1}^{m} \Pr\left[(X_j \leq h_j) \bigcap_{\substack{i=1\\ i\neq j}}^{m} (X_i < X_j h_i/h_j)\right] \tag{17}$$

(where 0/0 is interpreted as 1). Each term on the right-hand side of (17) can be expressed in terms of a multinormal integral involving only $(m-1)$ variables; e.g.,

$$\Pr\left[(X_m \leq h_m)\bigcap_{i=1}^{m-1}(X_i \leq X_m h_i/h_m)\right] = \int_{-\infty}^{h_m} \Pr\left[\bigcap_{i=1}^{m-1}(X_i \leq x h_i/h_m)\right] Z(x)\, dx. \tag{18}$$

By repeated application of (18), evaluation of $\Phi(\mathbf{h};\mathbf{R})$ can be made to depend on quadratures involving only univariate normal integrals. The process will become very cumbersome if m is not rather small ($m > 5$, say); the limitation on values of the h's should also be noted.

John [57] used a probabilistic argument to express integrals of the multinormal density in terms of integrals of multinormal densities with fewer variables.

The event $\bigcap_{j=1}^{m} (X_j \leq h_j)$ is equivalent to the event $\max_{1\leq j\leq m} (X_j - h_j) \leq 0$. Hence if $L \equiv L(h_1,\ldots,h_m) = \max_{1\leq j\leq m} (X_j - h_j)$,

$$(19) \qquad \Phi(\mathbf{h};\mathbf{R}) = \Pr\left[\bigcap_{j=1}^{m} (X_j \leq h_j)\right]$$

$$= \Pr[L \leq 0]$$

$$= \sum_{j=1}^{m} \int_{-\infty}^{0} \Pr\left[\bigcap_{\substack{i=1 \\ i\neq j}}^{m} (X_i < h_i) \mid X_j = h_j + t\right] Z(t + h_j)\, dt.$$

These methods, as applied in the special cases $m = 2, m = 3$ are discussed further in Chapter 36.

In order to render calculations simpler, Marsaglia [82] has utilized the relationship

$$\Phi(\mathbf{h};\mathbf{A} + \mathbf{R}) = E[\Phi(\mathbf{h} - \mathbf{Y};\mathbf{R})],$$

where $\mathbf{Y}$ have a joint multinormal distribution with variance covariance matrix $\mathbf{A}$, and expected value vector $\mathbf{O}$. (We have already noted in equation (13) that the definition of $\Phi(\cdot)$ also applies when $\mathbf{R}$ is a variance covariance matrix.) Choice of $\mathbf{A}$ and $\mathbf{R}$ has been discussed by Anderson [6].

4.2. *Orthant Probabilities*

The problem of evaluating $\Phi(\mathbf{h};\mathbf{R})$ may be specialized with respect to either $\mathbf{h}$ or $\mathbf{R}$, or both. If we take $\mathbf{h} = \mathbf{O}$, we have the problem of evaluating orthant probabilities. Since $\mathbf{R}$ is unchanged if each X_j is replaced by $-X_j$, we have

$$(20) \qquad \Phi_m(\mathbf{O};\mathbf{R}) = \frac{|\mathbf{R}|^{1/2}}{(2\pi)^{m/2}} \int_0^\infty \int_0^\infty \cdots \int_0^\infty \exp[-\tfrac{1}{2}\mathbf{x}'\mathbf{R}^{-1}\mathbf{x}]\, d\mathbf{x}.$$

The integral (20) was studied as early as 1858, when Schläfli [105] obtained a differential equation, corresponding to (15) with $\mathbf{h} = \mathbf{O}$.

A direct method of calculation can be based on an expansion of the ratio of the density function (12) to the density function with $\mathbf{R} = \mathbf{I}$. This was obtained by Mehler [85] in 1866 for bivariate distributions, and generalized by Kibble [69] in 1945 to the multinormal case. The expansion is

$$(21) \qquad Z_m(x;\mathbf{O};\mathbf{R}) = Z_m(x;\mathbf{O};\mathbf{I}) \sum_{j=0}^{\infty} (j!)^{-1} \sum_i{}^* C_i \prod_{\alpha\beta} \rho_{\alpha\beta} \prod_{t=1}^{m} H_{i_t}(x_t)$$

where $\sum^*$ is a sum over all possible sets of j $\rho_{\alpha\beta}$'s (including repeated values),

C_i = the number of different permutations of the $\rho_{\alpha\beta}$'s, i.e., $j!\left(\prod_k j_k!\right)^{-1}$ where the same ρ is repeated $j_1, j_2, \ldots, j_k$ times $\left(\sum_k j_k = j\right)$ $(\alpha < \beta)$,

i_t = the number of times t occurs among the suffices α,β in the ith term of $\sum^*$, and

$H_r(x)$ = the rth Hermite polynomial (Chapter 1, Section 3).

(A relatively simple derivation of (21) for the case $m = 2$ is given by Brown [19].)

Since each term of (21) is a product of functions of the form

$$\text{constant} \times H_l(x_l)e^{-(1/2)x_l^2},$$

it is possible to integrate term-by-term and so obtain a series expansion for $\Phi(\mathbf{O};\mathbf{R})$.

Kendall [62] has obtained an equivalent series expansion by working with the inversion formula for the density in terms of the characteristic function. From (4) we have

$$Z_m(\mathbf{x};\mathbf{R}) = (2\pi)^{-m/2}\int_{-\infty}^{\infty}\cdots\int_{-\infty}^{\infty}\exp(-i\mathbf{t}'\mathbf{x} - \tfrac{1}{2}\mathbf{t}'\mathbf{R}^{-1}\mathbf{t})\,d\mathbf{t}.$$

Kendall used the formula $(\alpha < \beta)$

$$(22)\qquad \exp(-\tfrac{1}{2}\mathbf{t}'\mathbf{R}^{-1}\mathbf{t}) = \exp(-\tfrac{1}{2}\mathbf{t}'\mathbf{t})\sum_{j=0}^{\infty}(-1)^j\sum{}^*\prod_{k=1}^{m}t_k^{j_k}\prod_{\alpha,\beta}(\rho_{\alpha\beta}^{j_{\alpha\beta}}/j_{\alpha\beta}!),$$

where $\sum^*$ now denotes summation over all $\{j_{\alpha,\beta}\}$ for which

$$\sum_{\alpha=1}^{m} j_\alpha = 1; \qquad \sum_\beta (j_{\alpha\beta} + j_{\beta\alpha}) = j_\alpha.$$

This gives, after some reduction,

$$(23)\qquad \Phi_m(\mathbf{O};\mathbf{R}) = \int_{-\infty}^{\infty}\cdots\int_{-\infty}^{\infty} Z_m(\mathbf{x};\mathbf{R})\,d\mathbf{x}$$

$$= \sum_{j=0}^{\infty}(-1)^j\sum{}^*\left(\prod_{k=1}^{m}A_{jk}\right)\prod_{\alpha,\beta}(\rho_{\alpha,\beta}^{j_{\alpha\beta}}/j_{\alpha\beta}!)$$

where

$$A_t = \begin{cases} \tfrac{1}{2} & \text{if } t = 0, \\ 0 & \text{if } t \text{ is even}, \\ \dfrac{1}{i\sqrt{2\pi}}\,\dfrac{(t-1)!}{2^{\frac{1}{2}(t-1)}[\frac{1}{2}(t-1)]!} & \text{if } t \text{ is odd}. \end{cases}$$

[Note that since each $\rho_{\alpha\beta}$ is counted twice in j, j must be even. This ensures that $\prod_{k=1}^{m} A_{j_k}$ must be real. It also means that $(-1)^j$ can be omitted from (23).]

Unfortunately, these series converge very slowly unless all the ρ_{ij}'s are small. An approximate formula is presented in Section 4.4.

We now consider results that can be obtained by giving **R** special forms.

4.3. *Some Special Cases*

The matrix **R** may be specialized in a number of ways. Ihm [52] has obtained a general formula for $\Pr[(X_1,X_2,\ldots,X_m) \text{ in } \Omega]$ which applies when the *variance-covariance* matrix is of form $\boldsymbol{\Delta} + c^2\mathbf{1}\mathbf{1}'$ where Δ is a positive definite diagonal matrix and $\mathbf{1}' = (1,1,\ldots,1)$. This means that

$$\text{(24)} \qquad \begin{aligned} \operatorname{var}(X_j) &= \delta_{jj} + c^2, \\ \operatorname{cov}(X_i,X_j) &= c^2. \end{aligned}$$

Ihm showed that, if $E[\mathbf{X}'] = (0,0,\ldots,0)$,

$$\text{(25)} \quad \Pr[(X_1,X_2,\ldots,X_m) \text{ in } \Omega] = \frac{c}{\sqrt{2\pi}} \int_{-\infty}^{\infty} e^{-(1/2)c^2t^2} \int \cdots_{\Omega} \int [(2\pi)^{m/2} |\Delta|^{1/2}]^{-1} \times \exp\left[-\tfrac{1}{2}\sum_{j=1}^{m} \delta_{jj}^{-1}(y_j - t)^2\right] d\mathbf{y}\, dt.$$

Although this is a multiple integral of $(m + 1)$th order, which is greater than the order (m) of the original multinormal integral, the integral is in general of simpler form. If the correlations can be expressed in the form $\rho_{ij} = \lambda_i\lambda_j$, for all i and j, then $X_1,X_2,\ldots,X_m$ can be represented as

$$X_j = \lambda_j U_0 + \sqrt{1 - \lambda_j}\, U_j \qquad (j = 1,2,\ldots,m),$$

where $U_0,U_1,\ldots,U_m$ are independent unit normal variables. This representation greatly facilitates calculation of probabilities. The inequality $(X_j \le h_j)$ is equivalent to

$$U_j \le (h_j - \lambda_j U_0)/\sqrt{1 - \lambda_j},$$

hence

$$\text{(26)} \qquad \Pr\left[\bigcap_{j=1}^{m} (X_j \le h_j)\right] = \int_{-\infty}^{\infty} Z(u_0) \prod_{j=1}^{m} \Phi\left(\frac{h_j - \lambda_j u_0}{\sqrt{(1 - \lambda_j)}}\right) du_0$$

(Dunnett and Sobel [34]).

For the special case $\rho_{ij} = \lambda_i/\lambda_j$, for all $i \leq j$, Curnow and Dunnett [24] have found reduction formulas for $m = 3, 4, 5$.

If all the correlations are equal and positive ($\rho_{ij} = \rho > 0$ for all i,j), then we have the representation

$$X_j = \sqrt{\rho}U_0 + \sqrt{1-\rho}U_j \qquad (j = 1,2,\ldots,m) \tag{27}$$

obtained by putting $\lambda_j = \sqrt{\rho}$. The inequality $X_j \leq h_j$ is equivalent to $U_j \leq (h_j - \sqrt{\rho}U_0)/\sqrt{1-\rho}$, and

$$\Pr\left[\bigcap_{j=1}^{m}(X_j \leq h_j)\right] = \int_{-\infty}^{\infty} Z(u_0)\prod_{j=1}^{m}\Phi\left(\frac{h_j - \sqrt{\rho}u_0}{\sqrt{1-\rho}}\right)du_0. \tag{28}$$

In the general case this must still be evaluated by numerical quadrature, but the reduction to a single integral makes the calculation much simpler. If also $h_1 = h_2 = \cdots = h_m = h$, we have

$$\Pr[\max(X_1,\ldots,X_m) \leq h] = \int_{-\infty}^{\infty} Z(u_0)\left[\Phi\left(\frac{h - \sqrt{\rho}u_0}{\sqrt{1-\rho}}\right)\right]^m du_0. \tag{29}$$

This formula has been obtained in a number of equivalent forms by Das [25], Dunnett [32], Dunnett and Sobel [34], Gupta [47], Ihm [52], Moran [90], Ruben [100], and Stuart [117], among others. Steck and Owen [116] have shown that this formula is valid for negative ρ as well, even though the integrand on the right-hand side is complex. These authors also obtain a useful recurrence relation (valid for ρ positive or negative). Denoting the probability in (25) by $F(h \mid \rho,m)$ they show that

$$F(h \mid \rho,m) = \sum_{j=1}^{m}(-1)^{j+1}\binom{m}{j}F(\alpha h \mid \rho',j)F(h \mid \rho,m-j), \tag{30}$$

where

$$\alpha = \left[\frac{1-\rho}{\{1+(m-1)\rho\}\{1+(m-2)\rho\}}\right]^{1/2}$$

and

$$\rho' = -\rho\{1+(m-2)\rho\}^{-1}.$$

When $\rho_{ij} = \rho$ for all i,j *and* $h = 0$ a number of simplifications are possible. We denote $\Phi_m(\mathbf{O};\mathbf{R})$ in this case by $L_m(\rho)$ for convenience.

Sampford (quoted by Moran [90]) showed that if $\rho > 0$, then

$$L_m(\rho) = \frac{1}{\sqrt{\pi}}\int_{-\infty}^{\infty} e^{-t^2}[1 - \Phi(at)]^m\,dt, \tag{31}$$

where $a = 2\rho/(1-\rho)$. Although the integral must be evaluated by quadrature, accurate values are easily obtained from simple summation formulas.

From Plackett's formula (15), putting $\rho_{ij} = \rho$ for all i,j and adding, we obtain

$$\frac{\partial L_m(\rho)}{\partial \rho} = \frac{m(m-1)}{4\pi(1-\rho^2)^{1/2}} L_{m-2}\left(\frac{\rho}{1+2\rho}\right), \tag{32}$$

from which (noting that $L_m(0) = 2^{-m}$), we have

$$L_m(\rho) = (\tfrac{1}{2})^m + \frac{m(m-1)}{4\pi} \cdot \int_0^r L_{m-2}\left(\frac{r}{1+2r}\right)(1-r^2)^{-1/2}\, dr \tag{33}$$

(Ruben [100]). Using the known values of $L_2(\rho)$ and $L_3(\rho)$ (see Chapter 36, Section 4) we find

(33)′

$$L_m(\rho) = 2^{-m}\left[1 + \frac{m^{(2)}}{\pi}\sin^{-1}\rho + \frac{m^{(4)}}{\pi^2}\int_0^\rho \frac{\sin^{-1}[r_1/(1+2r_1)]}{(1-r_1^2)^{1/2}}\, dr_1 + \cdots\right]$$

$$+ \frac{m^{(6)}}{\pi^3}\int_0^\rho\int_0^{r_2/(1+2r_2)} \frac{\sin^{-1}[r_1/(1+2r_1)]}{(1-r_1^2)^{1/2}} \frac{dr_1\, dr_2}{(1-r_2^2)^{1/2}} + \cdots\Bigg].$$

The $(j+1)$th term in the series on the right is

$$\frac{m^{(2j)}}{\pi^j} I_j(\rho),$$

where

$$I_j(\rho) = \int_0^\rho \int_0^{r_j/(1+2r_j)} \cdots \int_0^{r_2/(1+2r_2)} \frac{\sin^{-1}[r_1/(1+2r_1)]}{(1-r_1^2)^{1/2}} \frac{dr_1\, dr_2 \cdots dr_j}{\prod_{i=2}^{j}(1-r_i^2)^{1/2}}.$$

Bacon [10] gives a table of values of $I_2(\rho)$, $I_3(\rho)$, and $I_4(\rho)$ (Table 1).

David and Six [28] have shown that when $\rho_{ij} = \frac{1}{2}$ for all i,j then

$$\Pr\left[\bigcap_{t=1}^{u} (X_t \le 0) \bigcap_{t=u+1}^{m} (X_t > 0)\right] = (m+1)^{-1} \qquad \text{for} \quad u = 0,1,\ldots,m. \tag{34}$$

They also give tables of the probability that u or fewer of $X_1, X_2, \ldots, X_m$ are positive when $\rho_{ij} = \rho$, to three decimal places for $\rho = 0.4(0.025)0.5$; $m = 12, 14, 16, 20, 24, 36, 48, 96$ for various values of u.

Das [25] reduced the evaluation of L to an integral of the density function of $m + k$ *independent* normal variables, where k need not exceed m – (multiplicity of smallest eigenvalue of $\mathbf{R}$) (Marsaglia [82]).

To perform the reduction it is necessary to express $\mathbf{R}$ in the form

$$\mathbf{R} = c^2\mathbf{I}_m + \mathbf{BB}', \tag{35}$$

TABLE 1

Values of the Integrals $I_i(\rho)$, $i = 2,3,4$, Used in Evaluation of Multinormal Probabilities

ρ	$I_2(\rho)$	$I_3(\rho)$	$I_4(\rho)$
.00	.000000	.000000	.000000
.05	.001172	.000017	.000002
.10	.004404	.000117	.000022
.15	.009477	.000343	.000083
.20	.016067	.000712	.000203
.25	.024057	.001232	.000397
.30	.033375	.001907	.000673
.35	.043812	.002727	.001037
.40	.055459	.003706	.001495
.45	.068254	.004843	.002052
.50	.082247	.006152	.002714
.55	.097454	.007628	.003486
.60	.114012	.009291	.004379
.65	.132053	.011154	.005404
.70	.151813	.013243	.006580
.75	.173640	.015607	.007934
.80	.198120	.018306	.009506
.85	.226180	.021464	.011370
.90	.259820	.025310	.013668
.95	.303950	.030429	.016770
1.00	.411234	.043064	.024159

where $c > 0$ and $\mathbf{B}$ is a real $m \times k$ matrix. If (35) holds, then $X_1,X_2,\ldots,X_m$ can be represented by

$$c(Y_1,\ldots,Y_m) - (Z_1,\ldots,Z_k)\mathbf{B}'$$

with $Y_1,\ldots,Y_m, Z_1,\ldots,Z_k$ independent unit normal variables. Hence

$$\Pr\left[\bigcap_{j=1}^{m}(X_j < h_j)\right]$$

$$= \Pr\left[\bigcap_{j=1}^{m}\left(Y_j \leq \left\{h_j + \sum_{i=1}^{k} b_{ji}Z_i\right\}c^{-1}\right)\right]$$

$$= (2\pi)^{-(1/2)k}\int_{-\infty}^{\infty}\cdots\int_{-\infty}^{\infty} e^{-\frac{1}{2}\sum_{1}^{k} z_i^2}\prod_{j=1}^{m}\Phi\left(c^{-1}\left\{h_j + \sum_{i=1}^{k} b_{ji}z_i\right\}\right)dz_1\cdots dz_k.$$

This is a k-fold integral, so that it is very desirable to make k as small as possible.